T0306114

FACTOR SEPARATION IN THE ATMOSPHERE

Applications and Future Prospects

Modeling atmospheric processes in order to forecast the weather or future climate change is an extremely complex and computationally intensive undertaking. One of the main difficulties is that there are a huge number of factors that need to be taken into account, some of which are still poorly understood. The Alpert–Stein Factor Separation (FS) Methodology is a computational procedure that helps deal with these nonlinear factors. Pinhas Alpert was the main pioneer of the FS method in meteorology, and in recent years many scientists have applied this methodology to a range of modeling problems, including paleoclimatology, limnology, regional climate change, rainfall analysis, cloud modeling, pollution, crop growth, and other forecasting applications. This book is the first to describe the fundamentals of the method, and to bring together its many applications in the atmospheric sciences, with chapters from many of the leading atmospheric modeling teams around the world. The main audience is researchers and graduate students using the FS method, but it is also of interest to advanced students, researchers, and professionals across the atmospheric sciences.

PINHAS ALPERT is Professor of Dynamic Meteorology and Climate and Head of the Porter School of Environmental Sciences, Tel Aviv University, Israel. He is a co-author of more than 180 peer-reviewed articles mainly on aspects of mesoscale dynamics and climate. His research focuses on atmospheric dynamics, climatology, numerical methods, limited area modeling, and climate change.

TATIANA SHOLOKHMAN is currently a Ph.D. student at the Department of Geophysics and Planetary Science at Tel Aviv University, Israel.

FACTOR SEPARATION IN THE ATMOSPHERE

ATMOSPHERE

Applications and Future Prospects

Edited by

PINHAS ALPERT AND TATIANA SHOLOKHMAN
Tel Aviv University, Israel

CAMBRIDGE
UNIVERSITY PRESS

University Printing House, Cambridge CB2 8BS, United Kingdom

One Liberty Plaza, 20th Floor, New York, NY 10006, USA

477 Williamstown Road, Port Melbourne, VIC 3207, Australia

314-321, 3rd Floor, Plot 3, Splendor Forum, Jasola District Centre, New Delhi - 110025, India

103 Penang Road, #05-06/07, Visioncrest Commercial, Singapore 238467

Cambridge University Press is part of the University of Cambridge.

It furthers the University's mission by disseminating knowledge in the pursuit of
education, learning and research at the highest international levels of excellence.

www.cambridge.org
Information on this title: www.cambridge.org/9780521191739

© Cambridge University Press 2011

First published 2011

A catalogue record for this publication is available from the British Library

Library of Congress Cataloging in Publication data
Factor Separation in the Atmosphere : Applications and Future Prospects / edited by
Pinhas Alpert and Tatiana Sholokhman.
p.   cm.
Includes bibliographical references and index.
ISBN 978-0-521-19173-9
1. Atmospheric diffusion–Mathematical models.   2. Meteorology–Mathematical methods.
3. Factorization (Mathematics)   I. Alpert, Pinhas, 1949–   II. Sholokhman, Tatiana.
III. Title.
QC880.4.D44F33   2011
551.5–dc22   2010045705

ISBN 978-0-521-19173-9 Hardback

Contents

The color plates are situated between pages 110 and 111.

Contributors

Pinhas Alpert
Porter School of Environmental Sciences,
Tel Aviv University,
Israel

Adriana Beltrán-Przekurat
Department of Atmospheric and Oceanic Sciences and
Cooperative Institute for Research in Environmental Sciences,
216 UCB,
University of Colorado,
Boulder, CO 80309,
USA

André Berger
Institute of Astronomy and Geophysics G. Lemaître,
Université Catholique de Louvain,
Louvain-la-Neuve,
Belgium

F. Booker
USDA-ARS Plant Science Research Unit, and
Department of Crop Sciences,
North Carolina State University,
Raleigh, NC 27695,
USA

Martin Claussen
Max Planck Institute for Meteorology,
Bundesstr. 53,
20146 Hamburg,
Germany

William R. Cotton
Department of Atmospheric Science,
Colorado State University,
1371 Campus Delivery,
Fort Collins, CO 80523-1371,
USA

Eric Deleersnijder
Centre for Systems Engineering and
Applied Mechanics (CESAME),
Université Catholique de Louvain,
4 Avenue Georges Lemaître,
B-1348 Louvain-la-Neuve,
Belgium

Joseph L. Eastman
WindLogics Inc.,
201 4th St NW,
Grand Rapids, MN 55744,
USA

Olivier Gourgue
Centre for Systems Engineering and
Applied Mechanics (CESAME),
Université Catholique de Louvain,
4 Avenue Georges Lemaître,
B-1348 Louvain-la-Neuve,
Belgium

Joshua P. Hacker
National Center for Atmospheric Research,
Boulder, CO,
USA

T. N. Krishnamurti
Department of Meteorology,
Florida State University,
Tallahassee, FL 32306,
USA

Vinay Kumar
Department of Meteorology,
Florida State University,
Tallahassee, FL 32306
USA

Vincent Legat
Centre for Systems Engineering and
Applied Mechanics (CESAME),
Université Catholique de Louvain,
4 Avenue Georges Lemaître,
B-1348 Louvain-la-Neuve,
Belgium

Ming Lei
Department of Agronomy, and
Department of Earth and Atmospheric Sciences,
Purdue University,
Lilly Hall of Life Sciences,
915 W. State Street,
West Lafayette, IN 47907–2054,
USA

Emmanuel Marchal
N-Side s.a.
6 Chemin du Cyclotron,
B-1348 Louvain-la-Neuve,
Belgium

R. Mera
Department of Marine, Earth, and
Atmospheric Sciences,
North Carolina State University,
Raleigh, NC 27695,
USA

Gemma T. Narisma
Physics Department,
Ateneo de Manila University,
Loyola Heights,
Quezon City, 1108,
Philippines

Dev Niyogi
Department of Agronomy, and
Department of Earth and Atmospheric Sciences,
Purdue University,
Lilly Hall of Life Sciences,
915 W. State Street,
West Lafayette, IN 47907–2054,
USA

Roger A. Pielke Sr.
Department of Atmospheric and Oceanic Sciences and
Cooperative Institute for Research in Environmental Sciences,
216 UCB,
University of Colorado,
Boulder, CO 80309,
USA

A. J. Pitman
Climate Change Research Centre,
University of New South Wales,
Sydney, NSW, 2052,
Australia

Gerhard Reuter
Department of Earth and Atmospheric Sciences,
1–26 Earth Sciences Building,
University of Alberta,
Edmonton,
Alberta, T6G 2E3
Canada

R. Romero
Grup de Meteorologia,
Departament de Física,
Universitat de les Illes Balears,
Palma de Mallorca,
Spain

Dorita Rostkier-Edelstein
Israel Institute for Biological Research,
Ness-Ziona,
Israel

Christopher Rozoff
University of Wisconsin-Madison/CIMSS,
1225 W. Dayton St.,
Madison, WI 53706,
USA

Tatiana Sholokhman
Department of Geophysics and Planetary Science,
Tel Aviv University,
Israel

Susan C. van den Heever
Department of Atmospheric Science,
Colorado State University,
1371 Campus Delivery,
Fort Collins, CO 80523–1371,
USA

Laurent White
Geophysical Fluid Dynamics Laboratory,
Princeton University,
201 Forrestal Road,
Princeton, NJ 08536,
USA

G. Wilkerson
Department of Crop Sciences,
North Carolina State University,
Raleigh, NC 27695,
USA

Yongkang Xue
Department of Geography and
Department of Atmospheric Sciences,
University of California at Los Angeles,
Los Angeles, CA,
USA

Qiuzhen Yin
Institute of Astronomy and Geophysics G. Lemaître,
Université Catholique de Louvain,
Louvain-la-Neuve,
Belgium

Foreword

The Factor Separation method, pioneered in the now classic Stein and Alpert (1993) and Alpert *et al.* (1995) papers, provides a powerful, much-needed tool to assess both linear and nonlinear relationships among weather and climate forcings and feedbacks. As summarized in Chapter 1 of the book:

The FS method provides the methodology to distinguish between the pure influence of each and every factor as well as their mutual influence or synergies, which come into play when several factors, at least two, are "switched on" together. The understanding of which factor, or what combination of factors is most significant for the final result, is often very interesting in atmospheric studies. Discovering the most dominant factors in a specific problem can guide us to the important physical mechanisms and also to potential improvements in the model formulations.

Before this analysis procedure was introduced, numerical models usually performed sensitivity studies by turning on one forcing at a time, and used these results to decide what are the most important factors affecting a particular model simulation. However, we now recognize that such a linear type of analysis is incomplete and can even lead to the incorrect answer, as is illustrated in several chapters in this book.

This book provides a range of examples that illustrate the power of this analysis methodology for a range of spatial and temporal scales. The next step, besides applying the Alpert–Stein Factor Separation Methodology to additional atmospheric studies, should be to broaden it to include other geophysical disciplines.

Roger A. Pielke Sr.

Preface

This book is the result of almost two decades of research initiated in 1991 by an idea that sensitivity studies in the atmosphere were not being performed in the proper manner. My Ph.D. student Uri Stein was simulating the effects of two central factors on the rainfall over the Eastern Mediterranean with the Mesoscale Model MM4: Mediterranean Sea fluxes and topography. I called Uri, on a Friday in 1991, suggesting that the three simulations he was performing were not enough to capture the potential synergy or interaction between these two factors and that an additional simulation should be performed. Consequently, we developed the Factor Separation Methodology that allows for the separation of four potential contributions in our two-factor problem and 2^n simulations for any n-factor problem. This separation included specifically the double synergy term which is the net result of interaction between the two factors. As I had expected, we found immediately that the synergy term plays a central role in the atmosphere, one that is often larger than the net contribution of any singular factor's contribution.

When our paper was submitted in 1992 to the *Journal of Atmospheric Sciences*, I did not know what to expect. It seemed to me that the method was so basic that I could not believe we were the first to apply it in atmospheric sciences. However, our extensive literature search did not show any similar publications.

We were thrilled when the paper was accepted for publication, and I began to talk about it in a lecture series at several well-known institutions including the University of Oklahoma, where the lecture was attended by Doug Lilly and the late Tzvi Gal-Chen. I also spoke at CSU where the lecture was attended by Roger A. Pielke Sr. and Bill Cotton, and I presented the idea at the following four conferences over the years 1992–3:

- the International Workshop on Mediterranean Cyclone Studies, Trieste, Italy, 1992
- the Mesoscale Modeling Workshop, El-Paso, Texas, 1992
- the Yale Mintz Memorial Symposium, Jerusalem, Israel, 1992
- the U.S. Army Mesomet Panel Meeting, Monterey, CA, USA, 1993.

On all these occasions, the response was enthusiastic, which served to strengthen our faith in the great potential of our new methodology. I wish to particularly mention the outstandingly strong words of support from senior meteorologists including T. T. Warner (PSU, at that time), Don Johnson (Wisconsin), R. P. Pearce (Reading), D. Lilly (University of Oklahoma), R. A. Pielke Sr. (CSU), B. Cotton (CSU) A. Berger (Louvain-la-Neuve), T. N. Krishnamurti (FSU, Tallahassee University), T. Gal-Chen (University of Oklahoma) and late J. Neumann (Hebrew University, Jerusalem).

D. Lilly informed me that he decided immediately to apply our method to a turbulence study he was performing at that time with his student, L. Deng. In fact, Deng and Lilly presented their factor separation study in the same year (1992), even before our paper was published in 1993. Their reference is: L. Deng and D. K. Lilly (1992) Helicity effect on turbulent decay in a rotating frame, *10th Symposium on Turbulence and Diffusion*, Portland, OR, pp. 338–341, AMS.

This new application illustrated the relative ease of applying the new methodology to different applications – my presentation in Oklahoma was in May 1992 and their proceeding publication appeared just a few months later.

Meanwhile, we submitted more papers on the method including a study focusing on the mechanisms related to the Genoa cyclogenesis, which analysed four factors and required 16 simulations including one quadruple, four triple, and six double synergies. In 1995 we submitted a paper entitled, 'Synergism in weather and climate' by P. Alpert, U. Stein, M. Tsidulko, and B. U. Neeman. In response, we received the following beautiful and most encouraging words from Rainer Bleck, who was one of the referees:

Unless there are difficulties with this method that have yet to come to light and may limit its broad use, the synergistic analysis method developed by Stein and Alpert is likely to become an indispensable tool in our field. After reading this paper, very few investigators trying to isolate the effect of various physical factors on climate and circulation through numerical simulation will be able to argue that they are exempt from using this method.

I cannot think of any aspect in the paper requiring further work; in other words, the paper is essentially ready to be published in J. Climate as is. A few minor editorial suggestions are spelled out in the manuscript which I am returning to the editor.

This is one of the few cases where one waits to hear from the journal editor that the paper has been accepted, so that one can start referencing it in one's own work, talking about it in class, etc. (Rainer Bleck)

Interestingly, the paper was rejected by the Editor stating that it was not appropriate for that particular journal.

A few years after our initial efforts on factor separation, I established the Factor Separation Group email list which grew within a year or two to about 50 users all over the world; many of whom can be found in my acknowledgements list below because they have initiated and been part of many interesting discussions and developments over the years. The incorporation of the method into various and diverse atmospheric topics was quick, and some of those works provide the basis for the different applications in the following chapters of the present book.

A few words on the order of the chapters in the book are appropriate here. Following the introduction (Chapter 1) and the mathematical formulation of the method (Chapter 2), some analytical functions are analyzed (Chapter 3) based on the Master Thesis of Tatiana Sholokhman, my co-author of the present book. Following these initial chapters are eleven chapters on various applications with the general order decreasing from the macro-scale to the micro-scale. Therefore, the following chapters and applications are: Paleoclimate (Chapter 4), Mesometeorology (Chapter 5), Regional climate (Chapter 6), Heavy rainfall and Cyclogenesis (Chapter 7), Clouds (Chapter 8), Limnology (Chapter 9), Pollution (Chapter 10), Crop growth (Chapter 11), Sea breeze (Chapter 12) and two different forecasting applications (Chapters 13, 14). Chapter 15 discusses some difficulties and prospects of the methodology, including a comparison to a similar but different method applied mainly in biotechnology experiments. Chapter 14 by T. N. Krishnamurti also suggests and discusses a similar but different method for factor separation. Chapter 16 provides a summary. An important addition at the end is an Appendix, which lists, to the best of our knowledge, all articles and publications employing the Alpert–Stein Factor Separation Methodology at the time of publication. I wish to emphasize that up until the last moment we were informed of new publications using the method that we had not heard about before.

It should be noted that for consistency throughout the chapters of the book we have suggested that the method be referred to as the 'Alpert–Stein Methodology' or the 'Alpert–Stein Factor Separation Methodology'.

Special thanks go to the US–Israel BiNational Science Foundation (BSF) jointly with T. T. Warner (then at PSU), which funded our initial research on cyclogenesis over the Mediterranean that yielded the Factor Separation Methodology. Also, thanks go to the German–Israel Foundation (GIF), which continued the funding to our cyclogenesis study jointly with J. Egger (Munich University).

Cambridge University Press is to be congratulated on carefully bringing the book to its final stage with highly professional foresight and good advice.

I am indebted to a large number of colleagues and students for their interest and many suggestions through the years. In particular, I am very grateful to the following colleagues who contributed to the factor separation discussion during its various stages of evolution: Sergio Alonso, Andre Berger, Reiner Bleck, Bob Bornstein,

Itsik Carmona, Bill (W. R.) Cotton, Eric Deleersnijder, C. A. Doswell III, Lenny Druyan, Joe Egger, Hubert Gallee, the late Tzvi Gal-Chen, Víctor Homar, Agustí Jansá, Don Johnson, Alexander Khain, Simon (S. O.) Krichak, Krish (T. N.) Krishnamurti, Gad Levy, Doug (D. K.) Lilly, Barry Lynn, Benny U. Neeman, the late Jehuda Neumann, Dev Niyogi, Bob (R. P.) Pearce, Natalie Perlin, Roger (R. A.) Pielke, Sr. Clemente Ramis, Romu (R.) Romero, Dorita Rostkier-Edelstein, Moti Segal, Uri Stein, Dave (D. J.) Stensrud, Alan Thorpe, Marina Tsidulko, and Tom (T. T.) Warner.

Pinhas Alpert

1

Introduction

P. Alpert

1.1 Background

Numerical models provide a powerful tool for atmospheric research. One of the most common ways of utilizing a model is by performing sensitivity experiments. Their purpose is to isolate the effects of different factors on certain atmospheric fields in one or more case studies. Factors that have been tested in sensitivity studies include, for example, surface sensible and latent heat fluxes, latent heat release, horizontal and vertical resolution, sea surface temperatures, horizontal diffusion, surface stress, initial and boundary conditions, topography, surface moisture, atmospheric stability, and radiation. Sensitivity studies are performed either with real-data case studies or with idealized atmospheric situations.

Sensitivity studies often evaluate the influence of only one factor such as topography (Tibaldi *et al.*, 1980; Dell'Osso, 1984, McGinley and Goerss, 1986), but many investigations test several factors, and try to estimate their relative importance. One common method of evaluating the contribution of a specific factor is by analyzing the difference in fields between a control run and a simulation where this factor is switched off. The difference map is, in general, more illustrative than the presentation of the two individual simulations, and therefore has often been used (e.g., Tibaldi *et al.*, 1980; Mesinger and Strickler, 1982; Lannici *et al.*, 1987; Leslie *et al.*, 1987; Uccellini *et al.*, 1987; Mullen and Baumhafner, 1988; Kuo and Low-Nam, 1990). Presentation of a map showing the difference between two simulations is also a common procedure (Maddox *et al.*, 1981; Chang *et al.*, 1982, 1984; Chen and Cotton, 1983; Kenney and Smith, 1983; Alpert and Neumann, 1984; Benjamin and Carlson, 1986; McGinley and Goerss, 1986; Orlansky and Katzfey, 1987; Zack and Kaplan, 1987; Mailhot and Chouinard, 1989). It will be shown, however, that the difference map for two simulations, when more than one factor

Factor Separation in the Atmosphere: Applications and Future Prospects, ed. Pinhas Alpert and Tatiana Sholokhman. Published by Cambridge University Press. © Cambridge University Press 2011.

are involved, does not have a simple meaning, and in fact may be quite misleading. Suppose that the effects of two factors are investigated: the topography and the surface fluxes. Three simulations are then performed (as in many of the aforementioned studies): CON, the control simulation; NOT, the no-terrain simulation; and NOF, the no-fluxes simulation. What is the meaning of the difference between the simulated fields of CON and NOT? It shows the effect of the topography, and also of the joint effect (interaction or synergy) of topography with fluxes, because both the terrain and its synergy effects vanish when the terrain is switched off. In the same way, the difference between CON and NOF includes the effects of both the fluxes and the interaction between fluxes and topography. If the interaction factor is not isolated, the difference maps CON – NOT and CON – NOF cannot then be simply interpreted, as is commonly attempted.

Although the aforementioned interactions between factors are usually neglected, their significant role in some cases has been pointed out (e.g., Uccellini *et al.*, 1987; Mailhot and Chouinard, 1989). To the best of our knowledge, no sensitivity studies had been proposed to isolate these interactions before our Factor Separation methodology was introduced in 1993. The method presented in this book with many applications shows a consistent and quite simple approach for isolating the resulting fields due to any interactions among factors, as well as that due to the pure factors, using linear combinations of a number of simulations.

1.2 What is the concept behind the Factor Separation method?

It is close to two decades since the publication of our basic methodology in "Factor Separation in Numerical Simulation" (henceforth FS) by U. Stein and P. Alpert (1993), and today this method provides a powerful tool for atmospheric research. A relatively large number of studies in the past two decades have been devoted to various applications of this method in atmospheric research.

The present book provides first a theoretical investigation of this methodology, and defines common principles and analytical explorations of the factors and their synergies. It should be noted that most of the studies employing the Alpert–Stein Factor Separation Methodology used the numerical modeling framework, which commonly does not allow an analytical investigation of the methodology. Hence, we present the results of the method application with simple mathematical functions that, at first, may have no clear physical meaning. From these results we create the basis for investigation of simple functions that describe complicated processes.

How does the method work, and what do we need it for? The next chapter (Chapter 2) provides the full mathematical description of the method, for which we propose a basic intuitive explanation. The starting point for investigating the problem is the inclusion of several factors that are assumed to influence the final result. It is important to note that both physical and geophysical values used in the model are often approximated, or they may change very quickly by the function of the chosen factors. For example, the albedo of the surface depends on the structure of the surface, and can change rapidly with the weather. Another consideration is that the temporal variation is very important for modeling the problem, but it adds complexity as it includes additional factors. The FS method can identify the most important factors and their combinations that henceforth are also referred to as *synergisms*. The synergism is defined in the dictionary as "the action of two or more factors to achieve an effect of which each is individually incapable."

An additional problem that the FS method addresses is the stability of the chosen factor in the problem. This means, how does the result react to small changes in the factors' values? The FS standard method works on the principle of "on–off". In other words, the factors are switched off (zeroed) one-by-one, and the intermediate result is investigated independently for each case. The case where all the chosen factors are switched off is the basic case, which obviously does not depend on the factors and their various combinations. The opposite case, with all of the factors switched "on", is the "control" or "full" result, which includes all the factors and their synergistic contributions. The FS method provides the methodology to distinguish between the pure influence of each and every factor as well as their mutual influence or synergies, which come into play when several factors, at least two, are "switched on" together. The understanding of which factor, or what combination of factors, is most significant for the final result, is often very interesting in atmospheric studies. Discovering the most dominant factors in a specific problem can guide us to the underlying physical mechanisms and to potential improvement in the model formulations.

1.3 The philosophy of synergy

What is the simple meaning of synergism? Let us take, for example, the following real-life situation. A worker needs to push a load along X meters. Let us assume that the push lasts time T. This time may not be equal to the time that two workers require to push the same load along double the distance, i.e., $2X$ meters. If the workers are in good coordination, i.e., positive synergy between them, they will push the load to a greater distance than $2X$ meters during the same period of time. But if they do not coordinate well, i.e., negative synergy, the distance will be less than $2X$ meters. In this case, each worker may be considered as a factor in the problem, and the result is the distance over which the load will be transported during the specific time T. Another example of negative synergy for the above workers could result from external limits, for instance, if the total distance is limited to $1.5X$ meters. Then, even if the two workers coordinate very well, the synergism must be negative by at least $0.5X$. This can correspond in the atmosphere to the very common case of saturation, where, for example, it is manifested as humidity saturation for rainfall. In other words, processes in life are often not linear, and the situation is similar or even more pronounced in the atmosphere. Probably there are no factors in the real atmosphere that are not correlated at all. This infinite chain of interactions among factors, or some small subset of them, is commonly investigated by scientists, and involves a finite number of assumptions.

A most attractive feature of the FS approach is the ability to quantify synergies or the interaction processes that were found to play a central role in many atmospheric processes, as found by several of the applications discussed in this book. There seems to be some basic psychological tendency in human thinking to present results linearly, with the hidden assumption that the synergies are small or can be ignored. Nonlinearities in the atmosphere, however, are often significant, and therefore need to be calculated and separated from the pure contributions of each factor, as shown in the many very different FS applications presented in this book.

Another attractive feature of the FS methodology is the fact that the system is closed in the sense that the sum of all contributions of the pure factors and their synergies always yields the result given by the control run, in which all factors are switched on.

In some of the following chapters, the synergies can be given a very clear physical meaning that also yields a much better understanding of the complex phenomena being investigated.

The next chapter presents the full mathematical derivation of the FS methodology.

2

The Factor Separation Methodology and the fractional approach

T. Sholokhman and P. Alpert

2.1 Following Stein and Alpert (1993) (SA, in brief)

2.1.1 The proposed method for two factors

The value of any predicted field f depends on the initial and boundary conditions, as well as the model itself. If a continuous change is made in any factor ψ, the resulting field f in general changes in a continuous manner as well. This can be mathematically formulated as follows.

Let the factor ψ be multiplied by a changing coefficient c so that

$$\psi(c) = c\psi \tag{2.1}$$

The resulting field f is a continuous function of c:

$$f = f(c) \tag{2.2}$$

so that $f(1)$ is the value of f in the control simulation, and $f(0)$ is the value of f in the simulation where the factor ψ is omitted. In the notation that follows, f_0 and f_1 are used for $f(0)$ and $f(1)$, respectively.

It is always possible to decompose any function $f(c)$ into a constant part, $\hat{f}_0$, that is independent of c, and a c-dependent component, $\hat{f}(c)$, such that $\hat{f}(0) = 0$.

In this simple example,

$$\hat{f}_0 = f_0 \tag{2.3}$$

and

$$\hat{f}(c) = f(c) - f_0 \tag{2.4}$$

Factor Separation in the Atmosphere: Applications and Future Prospects, ed. Pinhas Alpert and Tatiana Sholokhman. Published by Cambridge University Press. © Cambridge University Press 2011.

It is important to understand the meaning of $\hat{f}_0$ and $\hat{f}_1$, the latter being a short form for $\hat{f}(1)$. The term $\hat{f}_1$ represents that fraction of f that is induced by the factor ψ, while $\hat{f}_0$ is the remaining part that does not depend on factor ψ. In order to obtain $\hat{f}_0$ and $\hat{f}_1$, two simulations must be performed, one with the ψ factor included (control) that results in f_1, and the other with ψ excluded, for obtaining the result for f_0:

$$f_1 = \hat{f}_0 + \hat{f}_1 \tag{2.5}$$

$$f_0 = \hat{f}_0 \tag{2.6}$$

Solution of the above equations for $\hat{f}$ in terms of the output field f yields

$$\hat{f}_0 = f_0 \tag{2.7}$$

$$\hat{f}_1 = f_1 - f_0 \tag{2.8}$$

The last equation shows that subtraction of the field f_0 (factor ψ excluded) from field f_1 (control run) results in that part of f that is solely induced by the factor ψ. This is how the method works for a single factor, and exemplifies the more general rule that is developed below.

2.1.2 Generalization of the method for n factors

Let the field f depend on n factors ψ_i, where $i = 1, 2, \ldots, n$. Each factor is multiplied by a coefficient c_i, where

$$f = f(c_1, c_2, c_3, \ldots, c_n) \tag{2.9}$$

By way of a Taylor series expansion, the function f can be decomposed as follows:

$$f(c_1, c_2, \ldots, c_n) = \hat{f}_0 + \sum_{i=1}^{n} \hat{f}_i(c_i) + \sum_{i,j=1,2}^{n-1,n} \hat{f}_{ij}(c_i, c_j)$$
$$+ \sum_{i,j,k=1,2,3}^{n-2,n-1,n} \hat{f}_{ijk}(c_i, c_j, c_k) + \cdots + \hat{f}_{123\ldots n}(c_1, c_2, c_3, \ldots, c_n) \tag{2.10}$$

Here, $\sum_{i,j=1,2}^{n-1,n}$ is the sum of all sorted pairs, and $\sum_{i,j,k=1,2,3}^{n-2,n-1,n}$ is the sum of all sorted trios, and so on.

A more complicated case, a fractional approach to the factor separation method, where c_i can obtain values between 0 and 1, is discussed below, Section 2.2.

For the simple case, factor ψ_i is fully switched on or off as c_i changes from 1 to 0, respectively. Each function $\hat{f}_{ijk...}(c_i, c_j, c_k, ...)$ becomes identically zero if any of its variables c_i is zero. Employing a symbol system in which f_{ij} is the value of f in a simulation with $c_i = c_j = 1$, while all the rest of the coefficients are zero, and setting $c_i (i = 1, 2, ..., n)$ to either 1 or 0 in (2.10), yields

$$f_0 = f(0, 0, ..., 0) = \hat{f}_0 \tag{2.11}$$

$$f_i = \hat{f}_i + \hat{f}_0 \tag{2.12}$$

$$f_{ij} = \hat{f}_{ij} + \hat{f}_i + \hat{f}_j + \hat{f}_0 \tag{2.13}$$

$$f_{ijk} = \hat{f}_{ijk} + \hat{f}_{ij} + \hat{f}_{jk} + \hat{f}_{ik} + \hat{f}_i + \hat{f}_j + \hat{f}_k + \hat{f}_0 \tag{2.14}$$

$$f_{123...n} = \hat{f}_{123...n} + \cdots + \sum_{i,j,k=1,2,3}^{n-2,n-1,n} \hat{f}_{ijk} + \sum_{i,j=1,2}^{n-1,n} \hat{f}_{ij} + \sum_{i=1}^{n} \hat{f}_i + \hat{f}_0 \tag{2.15}$$

$f_{ij} = f_{ij}(1, 1)$, and the same applies for all other terms.

Equations (2.11)–(2.15) contain $\binom{n}{0}, \binom{n}{1}, \binom{n}{2}, ..., \binom{n}{n}$ equations respectively. Hence, Eqs. (2.11)–(2.15) contain a total of 2^n equations for 2^n unknowns $\hat{f}_0, \hat{f}_1, ..., \hat{f}_n, \hat{f}_{12}, ..., \hat{f}_{n-1,n}, ..., \hat{f}_{123...n}$. This set of equations is solved by a recursive elimination of $\hat{f}_i$ from (2.12), then $\hat{f}_{ij}$ from (2.13), and so forth. The general solution then becomes

$$\hat{f}_{i_1 i_2 i_3 ... i_l} = \sum_{m=0}^{l} (-1)^{l-m} \left[\sum_{j_1,j_2,j_3,...,j_m=i_1,i_2,i_3,...,i_m}^{i_{l-m+1},i_{l-m+2},...,i_l} f_{j_1 j_2 j_3 ... j_m} \right] \tag{2.16}$$

where the sum $\sum_{j_1,j_2,j_3,...,j_m=i_1,i_2,i_3,...,i_m}^{i_{l-m+1},i_{l-m+2},...,i_l}$ is over all groups of m sorted indices chosen from l indices $i_1, i_2, i_3, ..., i_l$, where $0 \leqslant l \leqslant n$.

Let us take the case of two factors and then three factors.

First case: two factors

run	factor 1	factor 2		
f_{12}	on	on	$= \hat{f}_0 + \hat{f}_1 + \hat{f}_2 + \hat{f}_{12}$	(2.17)
f_1	on	off	$= \hat{f}_0 + \hat{f}_1$	(2.18)
f_2	off	on	$= \hat{f}_0 + \hat{f}_2$	(2.19)
f_0	off	off	$= \hat{f}_0$	(2.20)

The solution of (2.17)–(2.20) yields the following four terms:

Unrelated to factors 1 and 2 $\hat{f}_0 = f_0$ $\qquad\qquad\qquad\qquad\qquad$ (2.21)

Induced by the factor 1 (independent of 2) $\hat{f}_1 = f_1 - f_0$ $\qquad\qquad$ (2.22)

Induced by the factor 2 (independent of 1) $\hat{f}_2 = f_2 - f_0$ $\qquad\qquad$ (2.23)

Induced by the synergism of factors 1 and 2 $\hat{f}_{12} = f_{12} - (f_1 + f_2) + f_0$ $\quad$ (2.24)

Second case: three factors

In the case of three factors, (2.16) yields eight (2^n) equations.

$$\hat{f}_0 = f_0 \qquad\qquad\qquad\qquad\qquad\qquad (2.25)$$

$$\hat{f}_1 = f_1 - f_0 \qquad\qquad\qquad\qquad\qquad (2.26)$$

$$\hat{f}_2 = f_2 - f_0 \qquad\qquad\qquad\qquad\qquad (2.27)$$

$$\hat{f}_3 = f_3 - f_0 \qquad\qquad\qquad\qquad\qquad (2.28)$$

$$\hat{f}_{12} = f_{12} - (f_1 + f_2) + f_0 \qquad\qquad\qquad (2.29)$$

$$\hat{f}_{13} = f_{13} - (f_1 + f_3) + f_0 \qquad\qquad\qquad (2.30)$$

$$\hat{f}_{23} = f_{23} - (f_2 + f_3) + f_0 \qquad\qquad\qquad (2.31)$$

$$\hat{f}_{123} = f_{123} - (f_{12} + f_{13} + f_{23}) + (f_1 + f_2 + f_3) - f_0 \qquad (2.32)$$

Hence, eight simulations are necessary with three factors.

The result is then not only the separation of the factors for $\hat{f}_1, \hat{f}_2, \hat{f}_3$, but also all the possible combinations of these factors, i.e. $\hat{f}_{12}, \hat{f}_{23}, \hat{f}_{13}$, and $\hat{f}_{123}$. The factor $\hat{f}_{123}$, for instance, is the contribution due to the triple interaction among the three factors under evaluation. Next, the method is analyzed in a study of the fractional approach.

2.2 The fractional approach: following Krichak and Alpert (2002) (KA, in brief)

Let the system of the model equations with both factors excluded be considered as a base system (BS). When the factors are included, the total time change of the model variable f at a particular point contains the contribution of each factor under consideration, as well as the contributions due to interactions among the factors. In addition, the f time change also contains contributions of the interactions of each factor being analyzed by the BS.

Let a and b be the two factors (physical effects) under deliberations 1 and 2, respectively.

In the case when both the a and b factors are taking part in the analysis:

$$f_0 = \hat{f}_0 \tag{2.33}$$

$$f_{01} = \hat{f}_0 + \hat{f}_1 + \hat{f}_{01} \tag{2.34}$$

$$f_{02} = \hat{f}_0 + \hat{f}_2 + \hat{f}_{02} \tag{2.35}$$

$$f_{012} = \hat{f}_0 + \hat{f}_1 + \hat{f}_2 + \hat{f}_{12} + \hat{f}_{01} + \hat{f}_{02} + \hat{f}_{012} \tag{2.36}$$

Here:

- f_0 is the time variation of a model characteristic f obtained by integration of the model with the chosen factors a and b excluded (BS);
- f_{01} is the time variation of a model characteristic f obtained by integration of the model with factor a included;
- f_{02} is the same, but for the b factor; and
- f_{012} is the same, but when both factors are included.

The $|\hat{\ }|$ terms of the $\hat{f}_\psi$ type represent fractions of f that are contributed by the ψ factor:

- $\hat{f}_{01}$ is the contribution of interaction of the factor a with the BS;
- $\hat{f}_{02}$ is the same, but for factor b;
- $\hat{f}_{012}$ is the contribution of the joint interaction of the two (a and b) factors with the BS;
- $\hat{f}_0$ is the pure contribution of the BS solution to the time variation of f.
- $\hat{f}_1$ is the pure contribution of the factor a to the time variation of f;
- $\hat{f}_2$ is the same as $\hat{f}_1$, but for factor b; and
- $\hat{f}_{12}$ is the pure contribution due to the a and b interaction – synergy a and b.

Solving the system (2.33)–(2.36) for the pure contributions $(\hat{f}_0, \hat{f}_1, \hat{f}_2, \hat{f}_{12})$ yields:

$$\hat{f}_0 = f_0 \tag{2.37}$$

$$\hat{f}_1 = f_{01} - f_0 - \hat{f}_{01} \tag{2.38}$$

$$\hat{f}_2 = f_{02} - f_0 - \hat{f}_{02} \tag{2.39}$$

$$\hat{f}_{12} = f_{012} - (f_{01} + f_{02}) + f_0 - \hat{f}_{012} \tag{2.40}$$

In many cases, the role of the BS-related contributions $(\hat{f}_{01}, \hat{f}_{02}, \hat{f}_{012})$ may be neglected. This allows using the standard FS formulation as suggested in the previous chapter. In this case, four model simulations are sufficient to evaluate the role of two chosen factors (a and b) and that of their synergic interaction. These are the simulations with the a and b factors excluded; the a factor is excluded, while b is included; the a factor is included, but b is excluded; and both a and b are included.

Thereby, after a demonstration of the methodology, we pass to mathematics. By applying the FS method to the simple mathematical functions, the trivial and basic rules are derived.

3

Investigation of the Factor Separation features for basic mathematical functions

T. Sholokhman and P. Alpert

3.1 FS for some fundamental functions

In spite of wide use of the FS method for complex functions, in order to investigate the FS method analytically, we start with only a few examples of mathematical functions chosen from a wide variety of functions. In other words, the FS method is checked out on simple mathematical functions that build up a base for meteorological or, in general, physical fields.

3.1.1 A bivariate polynomial

A bivariate polynomial is described by the function

$$f(x_1, x_2) = ax_1 + bx_2 + kx_1x_2 + d.$$

Let the factors x_i be multiplied by a varying coefficient c_i:

$$f(c_1, c_2) = ac_1x_1 + bc_2x_2 + kc_1c_2x_1x_2 + d \qquad (3.1)$$

It is probably preferable to write the function $f(x_1, x_2)$ as $f(c_1x_1, c_2x_2)$, but here we focus on the variable c_i (just simplified writing).

It can easily be seen that the values of f and $\hat{f}$ in the different simulations are related through:

$$
\begin{aligned}
f_0 &= d \\
f_1 &= ax_1 + d \\
f_2 &= bx_2 + d \\
f_{12} &= ax_1 + bx_2 + kx_1x_2 + d
\end{aligned}
\qquad (3.2)
$$

Factor Separation in the Atmosphere: Applications and Future Prospects, ed. Pinhas Alpert and Tatiana Sholokhman. Published by Cambridge University Press. © Cambridge University Press 2011.

$$\hat{f}_0 = d$$
$$\hat{f}_1 = ax_1$$
$$\hat{f}_2 = bx_2 \tag{3.3}$$
$$\hat{f}_{12} = kx_1x_2$$

This function is a form of Taylor series expansion by itself. Hence, this is a trivial case. The four different contributions are naturally expected since the factors' contributions are proportional to the corresponding variables, i.e., ax_1, bx_2, while the synergy is the term that includes both factors, i.e., kx_1x_2.

3.1.2 An exponential function

An exponential function $f(x_1, x_2) = e^{ax_1 + bx_2 + d}$ was chosen.

In the same way, the factors x_i are multiplied by a varying coefficient c_i and consequently we have

$$f(c_1, c_2) = e^{ac_1x_1 + bc_2x_2 + d} \tag{3.4}$$

The four simulations in this case yield

$$f_0 = e^d$$
$$f_1 = e^{ax_1 + d}$$
$$f_2 = e^{bx_2 + d} \tag{3.5}$$
$$f_{12} = e^{ax_1 + bx_2 + d}$$

And the four contributions are

$$\hat{f}_0 = e^d$$
$$\hat{f}_1 = e^{ax_1 + d} - e^d = e^d(e^{ax_1} - 1)$$
$$\hat{f}_2 = e^{bx_2 + d} - e^d = e^d(e^{bx_2} - 1) \tag{3.6}$$
$$\hat{f}_{12} = e^{ax_1 + bx_2 + d} - e^{ax_1 + d} - e^{bx_2 + d} + e^d = e^d(e^{ax_1 + bx_2} - e^{ax_1} - e^{bx_2} + 1)$$

To examine the consistency of the synergism term, the direct method using the Taylor series is applied.

The Taylor series for $f(c_1, c_2)$ in (3.4) about (0,0) is

$$f(c_1, c_2) = e^d + ae^d c_1 x_1 + be^d c_2 x_2 + \frac{1}{2} a^2 e^d c_1^2 x_1^2 + \frac{1}{2} b^2 e^d c_2^2 x_2^2 + abe^d c_1 c_2 x_1 x_2$$

$$+ \frac{1}{6} a^3 e^d c_1^3 x_1^3 + \frac{1}{6} b^3 e^d c_2^3 x_2^3 + \frac{1}{2} a^2 be^d c_1^2 c_2 x_1^2 x_2 + \frac{1}{2} ab^2 e^d c_1 c_2^2 x_1 x_2^2$$

$$+ \frac{1}{24} a^4 e^d c_1^4 x_1^4 + \frac{1}{24} b^4 e^d c_2^4 x_2^4 + \frac{1}{4} a^2 b^2 e^d c_1^2 c_2^2 x_1^2 x_2^2$$

$$+ \frac{1}{6} a^3 be^d c_1^3 c_2 x_1^3 x_2 + \frac{1}{6} ab^3 e^d c_1 c_2^3 x_1 x_2^3 + \cdots \qquad (3.7)$$

and the simulations in the FS method are

$$c_1 = c_2 = 0 \Rightarrow f_0 = e^d$$

$$c_1 = 1, c_2 = 0 \Rightarrow f_1 = e^d + ae^d x_1 + \frac{1}{2} a^2 e^d x_1^2 + \frac{1}{6} a^3 e^d x_1^3 + \frac{1}{24} a^4 e^d x_1^4 + \cdots$$

$$c_1 = 0, c_2 = 1 \Rightarrow f_2 = e^d + be^d x_2 + \frac{1}{2} b^2 e^d x_2^2 + \frac{1}{6} b^3 e^d x_2^3 + \frac{1}{24} b^4 e^d x_2^4 + \cdots$$

$$c_1 = c_2 = 1 \Rightarrow f_{12} = e^d + ae^d x_1 + be^d x_2 + \frac{1}{2} a^2 e^d x_1^2 + \frac{1}{2} b^2 e^d x_2^2 + abe^d x_1 x_2$$

$$+ \frac{1}{6} a^3 e^d x_1^3 + \frac{1}{6} b^3 e^d x_2^3 + \frac{1}{2} a^2 be^d x_1^2 x_2 + \frac{1}{2} ab^2 e^d x_1 x_2^2 + \frac{1}{24} a^4 e^d x_1^4$$

$$+ \frac{1}{24} b^4 e^d x_2^4 + \frac{1}{4} a^2 b^2 e^d x_1^2 x_2^2 + \frac{1}{6} a^3 be^d x_1^3 x_2 + \frac{1}{6} ab^3 e^d x_1 x_2^3 + \cdots \qquad (3.8)$$

Hence, the synergism function becomes

$$\hat{f}_{12} = f_{12} - (f_1 + f_2) + f_0 = abe^d x_1 x_2 + \frac{1}{2} a^2 be^d x_1^2 x_2 + \frac{1}{2} ab^2 e^d x_1 x_2^2$$

$$+ \frac{1}{4} a^2 b^2 e^d x_1^2 x_2^2 + \frac{1}{6} a^3 be^d x_1^3 x_2 + \frac{1}{6} ab^3 e^d x_1 x_2^3 + \cdots \qquad (3.9)$$

For comparison, the Taylor series for $\hat{f}_{12}$ from (3.6) about (0, 0) is

$$\hat{f}_{12} = abe^d x_1 x_2 + \frac{1}{2} a^2 be^d x_1^2 x_2 + \frac{1}{2} ab^2 e^d x_1 x_2^2$$

$$+ \frac{1}{4} a^2 b^2 e^d x_1^2 x_2^2 + \frac{1}{6} a^3 be^d x_1^3 x_2 + \frac{1}{6} ab^3 e^d x_1 x_2^3 + \cdots \qquad (3.10)$$

As can be seen, the same result was obtained in the two different calculations, i.e., equations (3.9) and (3.10).

3.1.3 A rational function

A simple rational function is chosen as $f(x_1,x_2) = \frac{1}{ax_1+bx_2+d}$.

$$f(c_1,c_2) = \frac{1}{ac_1x_1 + bc_2x_2 + d} \tag{3.11}$$

The four functions of f and four functions of $\hat{f}$ are derived:

$$
\begin{aligned}
f_0 &= \frac{1}{d}\\
f_1 &= \frac{1}{ax_1+d}\\
f_2 &= \frac{1}{bx_2+d}\\
f_{12} &= \frac{1}{ax_1+bx_2+d}
\end{aligned}
\tag{3.12}
$$

and the corresponding contribution functions are

$$
\begin{aligned}
\hat{f}_0 &= \frac{1}{d}\\
\hat{f}_1 &= \frac{1}{ax_1+d} - \frac{1}{d}\\
\hat{f}_2 &= \frac{1}{bx_2+d} - \frac{1}{d}\\
\hat{f}_{12} &= \frac{1}{ax_1+bx_2+d} - \frac{1}{ax_1+d} - \frac{1}{bx_2+d} + \frac{1}{d}
\end{aligned}
\tag{3.13}
$$

First, the Taylor series for $f(c_1,c_2)$ about $(0,0)$ is derived:

$$
\begin{aligned}
f(c_1,c_2) =\ & \frac{1}{d} - \frac{a}{d^2}c_1x_1 - \frac{b}{d^2}c_2x_2 + \frac{a^2}{d^3}c_1^2x_1^2 + \frac{b^2}{d^3}c_2^2x_2^2 + 2\frac{ab}{d^3}c_1c_2x_1x_2\\
& - \frac{a^3}{d^4}c_1^3x_1^3 - \frac{b^3}{d^4}c_2^3x_2^3 - 3\frac{a^2b}{d^4}c_1^2c_2x_1^2x_2 - 3\frac{ab^2}{d^4}c_1c_2^2x_1x_2^2\\
& + \frac{a^4}{d^5}c_1^4x_1^4 + \frac{b^4}{d^5}c_2^4x_2^4 + 4\frac{a^3b}{d^5}c_1^3c_2x_1^3x_2 + 4\frac{ab^3}{d^5}c_1c_2^3x_1x_2^3\\
& + 6\frac{a^2b^2}{d^5}c_1^2c_2^2x_1^2x_2^2 + \cdots
\end{aligned}
\tag{3.14}
$$

and by the FS method:

$$c_1 = c_2 = 0 \Rightarrow f_0 = \frac{1}{d}$$

$$c_1 = 1,\ c_2 = 0 \Rightarrow f_1 = \frac{1}{d} - \frac{a}{d^2}x_1 + \frac{a^2}{d^3}x_1^2 - \frac{a^3}{d^4}x_1^3 + \frac{a^4}{d^5}x_1^4 + \cdots$$

$$c_1 = 0,\ c_2 = 1 \Rightarrow f_2 = \frac{1}{d} - \frac{b}{d^2}x_2 + \frac{b^2}{d^3}x_2^2 - \frac{b^3}{d^4}x_2^3 + \frac{b^4}{d^5}x_2^4 + \cdots$$

$$c_1 = c_2 = 1 \Rightarrow f_{12} = \frac{1}{d} - \frac{a}{d^2}x_1 - \frac{b}{d^2}x_2 + \frac{a^2}{d^3}x_1^2 + \frac{b^2}{d^3}x_2^2 + 2\frac{ab}{d^3}x_1x_2$$

$$- \frac{a^3}{d^4}x_1^3 - \frac{b^3}{d^4}x_2^3 - 3\frac{a^2b}{d^4}x_1^2x_2 - 3\frac{ab^2}{d^4}x_1x_2^2 + \frac{a^4}{d^5}x_1^4$$

$$+ \frac{b^4}{d^5}x_2^4 + 4\frac{a^3b}{d^5}x_1^3x_2 + 4\frac{ab^3}{d^5}x_1x_2^3 + 6\frac{a^2b^2}{d^5}x_1^2x_2^2 + \cdots$$

$$(3.15)$$

and the synergism, by the FS formula, becomes

$$\hat{f}_{12} = f_{12} - (f_1 + f_2) + f_0 = 2\frac{ab}{d^3}x_1x_2 - 3\frac{a^2b}{d^4}x_1^2x_2 - 3\frac{ab^2}{d^4}x_1x_2^2 + 4\frac{a^3b}{d^5}x_1^3x_2$$

$$+ 4\frac{ab^3}{d^5}x_1x_2^3 + 6\frac{a^2b^2}{d^5}x_1^2x_2^2 + \cdots \qquad (3.16)$$

Hence, using the second method, the Taylor series for $\hat{f}_{12}$ about $(0, 0)$ yields

$$\hat{f}_{12} = 2\frac{ab}{d^3}x_1x_2 - 3\frac{a^2b}{d^4}x_1^2x_2 - 3\frac{ab^2}{d^4}x_1x_2^2$$

$$+ 4\frac{a^3b}{d^5}x_1^3x_2 + 4\frac{ab^3}{d^5}x_1x_2^3 + 6\frac{a^2b^2}{d^5}x_1^2x_2^2 + \cdots \qquad (3.17)$$

And, as in the previous example, the same result is obtained in equations (3.16) and (3.17).

To sum up, this chapter proves the uniqueness of the synergism in the model for some simple functions. This means that it does not matter how the solution is derived; the same synergic interaction is obtained. These examples provide the basis for the next section, where the synergism's behavior is investigated.

3.2 Analytical investigation of the synergism

3.2.1 "Not interesting case": the rule of zero synergism

The zero-synergism rule is examined with the goal of studying an easy, simple case in order to better understand the nature of the synergism concept. In theory the zero synergy can be found in simple mathematical functions as shown in the next section, but in practice it could be difficult to identify the situation that includes the interaction of several factors that becomes zero. In realistic situations, however, only in cases with many simplifications may the synergy tend to be zero. On the other hand, the analysis of analytical functions, as performed here, could give a clearer picture for the zero-synergy cases.

Hence, the following fact is the result of the FS method:
The synergism exists only if the Taylor expansion has at least one term like

$$x_1^{i_1} x_2^{i_2} x_3^{i_3} \ldots x_n^{i_n} \text{ when at least two of } i_1 i_2 \ldots i_n \neq 0$$

Let the field f depend on two factors: $f(x_1, x_2)$
Let the factors x_1 and x_2 be multiplied by varying coefficients c_1 and c_2 so that

$$x_i(c_i) = c_i x_i, \quad 0 \leqslant c_i \leqslant 1, \quad i = 1, 2$$

The resulting field is a continuous function of $f = f(c_1, c_2)$.

The factor separation for two factors as has been shown is

$$
\begin{aligned}
f_0 &\to c_1 = c_2 = 0 \to \hat{f}_0 = f_0 \\
f_1 &\to c_1 = 1, \ c_2 = 0 \to \hat{f}_1 = f_1 - f_0 \\
f_2 &\to c_1 = 0, \ c_2 = 1 \to \hat{f}_2 = f_2 - f_0 \\
f_{12} &\to c_1 = 1, \ c_2 = 1 \to \hat{f}_{12} = f_{12} - (f_1 + f_2) - f_0
\end{aligned}
\tag{3.18}
$$

If all partial derivatives of the function f of two variables exist, then the Taylor expansion of f about the point (a, b) can be written

$$
\begin{aligned}
f(x_1, x_2) = f(x_1^0, x_2^0) &+ \left[\frac{\partial}{\partial x_1}(x_1 - x_1^0) + \frac{\partial}{\partial x_2}(x_2 - x_2^0) \right] f_{\substack{x_1 = x_1^0 \\ x_2 = x_2^0}} \\
&+ \frac{1}{2!} \left[\frac{\partial}{\partial x_1}(x_1 - x_1^0) + \frac{\partial}{\partial x_2}(x_2 - x_2^0) \right]^2 f_{\substack{x_1 = x_1^0 \\ x_2 = x_2^0}} + \cdots \\
&+ \frac{1}{n!} \left[\frac{\partial}{\partial x_1}(x_1 - x_1^0) + \frac{\partial}{\partial x_2}(x_2 - x_2^0) \right]^n f_{\substack{x_1 = x_1^0 \\ x_2 = x_2^0}} + \cdots
\end{aligned}
\tag{3.19}
$$

At the point $x_1^0 = 0, x_2^0 = 0$, the above expression simplifies to the following form:

$$f(x_1, x_2) = f(0,0) + \sum_{i=1} a_i x_1^i \left. \frac{\partial^i f}{\partial x_1^i} \right|_{(0,0)} + \sum_{i=1} b_i x_2^i \left. \frac{\partial^i f}{\partial x_2^i} \right|_{(0,0)}$$

$$+ \sum_{\substack{i+j \geqslant 2 \\ i,j \neq 0}} k_{ij} x_1^i x_2^j \left. \frac{\partial^{i+j} f}{\partial x_1^i \partial x_2^j} \right|_{(0,0)} \tag{3.20}$$

where a_i, b_i, and k_{ij} are rational coefficients from equation (3.19).

Equation (3.20) can be written as

$$f(c_1, c_2) = f(0,0) + c_1 \sum_{i=1} a_i x_1^i \left. \frac{\partial^i f}{\partial x_1^i} \right|_{(0,0)} + c_2 \sum_{i=1} b_i x_2^i \left. \frac{\partial^i f}{\partial x_2^i} \right|_{(0,0)}$$

$$+ c_1 c_2 \sum_{\substack{i+j \geqslant 2 \\ i,j \neq 0}} k_{ij} x_1^i x_2^j \left. \frac{\partial^{i+j} f}{\partial x_1^i \partial x_2^j} \right|_{(0,0)}$$

By switching on/off some given factors (c_1, c_2) in the numerical simulations, the role played by these factors in our problem, can be isolated. Equation (3.20) then yields

$$f_0 = f(0,0)$$

$$f_1 = f(0,0) + \sum_{i=1} a_i x_1^i \left. \frac{\partial^i f}{\partial x_1^i} \right|_{(0,0)}$$

$$f_2 = f(0,0) + \sum_{i=1} b_i x_2^i \left. \frac{\partial^i f}{\partial x_2^i} \right|_{(0,0)} \tag{3.21}$$

$$f_{12} = f(0,0) + \cdots = f(x_1, x_2)$$

The pure influences are therefore found by substituting (3.21) into (3.18):

$$\hat{f}_0 = f(0,0)$$

$$\hat{f}_1 = \sum_{i=1} a_i x_1^i \left. \frac{\partial^i f}{\partial x_1^i} \right|_{(0,0)}$$

$$\hat{f}_2 = \sum_{i=1} b_i x_2^i \left. \frac{\partial^i f}{\partial x_2^i} \right|_{(0,0)}$$

$$\hat{f}_{12} = \sum_{\substack{i+j \geq 2 \\ i,j \neq 0}} k_{ij} x_1^i x_2^j \left. \frac{\partial^{i+j} f}{\partial x_1^i \partial x_2^j} \right|_{(0,0)} \tag{3.22}$$

Hence, only pairs such as in $\hat{f}_{12}$ cause the existence of the synergism in this case.

Certainly in the more general case, it is not a "simple point" $(0,0)$. Also, the field f may depend on n factors. So this "not interesting" result yields the following rule:

If any field f can be written in the form

$$f(x_1, x_2 \ldots, x_n) = \sum_{i=1}^{n} f(x_i) \tag{3.23}$$

the synergism is equal to zero by definition.

3.2.2 Synergism for fundamental functions

Giving examples is a useful way of supporting the argument. Each of the following short solutions belongs to one of the three groups: (i) there is no synergic interaction, or in other words, the synergy term is zero; (ii) the case of interaction acting in opposition to the pure factors; and (iii) a dominant synergism. All of the above cases are of interest and may occur in the general study of the FS method.

(a) A bivariate polynomial $f(x_1, x_2) = a x_1 + b x_2 + k x_1 x_2 + d$

$$f(c_1, c_2) = a c_1 x_1 + b c_2 x_2 + k c_1 c_2 x_1 x_2 + d$$

$$
\begin{array}{ll}
f_0 = d & \hat{f}_0 = d \\
f_1 = a x_1 + d & \hat{f}_1 = a x_1 \\
f_2 = b x_2 + d & \hat{f}_2 = b x_2 \\
f_{12} = a x_1 + b x_2 + k x_1 x_2 + d & \hat{f}_{12} = k x_1 x_2
\end{array} \tag{3.24}
$$

(1) The synergism is zero

From equation (3.18), the synergism equals zero by definition when $k = 0$, or in the series of the points when one of coordinates, i.e. x_1, x_2 is zero.

(2) The opposed influence of the synergism

Here, the synergism works in the opposite direction to that of the resulting field; which means that

$$\text{either } \begin{cases} f_{12} > 0 \\ \hat{f}_{12} < 0 \end{cases} \quad \text{or} \quad \begin{cases} f_{12} < 0 \\ \hat{f}_{12} > 0 \end{cases} \tag{3.25}$$

Substituting the relevant functions from (3.24) into (3.25),

$$\text{either } \begin{cases} ax_1 + bx_2 + kx_1x_2 + d > 0 \\ kx_1x_2 < 0 \end{cases} \quad \text{or} \quad \begin{cases} ax_1 + bx_2 + kx_1x_2 + d < 0 \\ kx_1x_2 > 0 \end{cases} \tag{3.26}$$

There are many points that solve the inequalities of equation (3.26). These points define the field in which every factor by itself, i.e., its pure contribution, works "for" the final result, but together they work "against" it.

(3) The dominant synergism

This is the case when the synergism term is greater than that of any of the factors acting separately:

$$\begin{cases} f_{12} > 0 \\ \hat{f}_{12} > \hat{f}_1 \\ \hat{f}_{12} > \hat{f}_2 \end{cases} \quad \text{or} \quad \begin{cases} f_{12} < 0 \\ \hat{f}_{12} < \hat{f}_1 \\ \hat{f}_{12} < \hat{f}_2 \end{cases} \tag{3.27}$$

Substituting (3.24) into (3.27) yields

$$\begin{cases} ax_1 + bx_2 + kx_1x_2 + d > 0 \\ kx_1x_2 > ax_1 \\ kx_1x_2 > bx_2 \end{cases} \quad \text{or} \quad \begin{cases} ax_1 + bx_2 + kx_1x_2 + d < 0 \\ kx_1x_2 < ax_1 \\ kx_1x_2 < bx_2 \end{cases} \tag{3.28}$$

Transposition from right side to left in (3.28) leads to

$$\begin{cases} ax_1 + bx_2 + kx_1x_2 + d > 0 \\ -ax_1 + kx_1x_2 > 0 \\ -bx_2 + kx_1x_2 > 0 \end{cases} \quad \text{or} \quad \begin{cases} ax_1 + bx_2 + kx_1x_2 + d < 0 \\ -ax_1 + kx_1x_2 < 0 \\ -bx_2 + kx_1x_2 < 0 \end{cases} \tag{3.29}$$

Addition all of three inequalities, and the separate addition of the first inequality and second inequality, and the separate addition of the first and third inequalities results in

$$\begin{cases} kx_1x_2 > -\dfrac{d}{3} \\ kx_1x_2 > ax_1 \\ kx_1x_2 > bx_2 \end{cases} \quad \text{or} \quad \begin{cases} kx_1x_2 < -\dfrac{d}{3} \\ kx_1x_2 < ax_1 \\ kx_1x_2 < bx_2 \end{cases} \tag{3.30}$$

In the last two systems, the synergism controls the final result in some specific cases. For a given example, in this way it is possible to see a range where the factors contribute only a small percentage of the result, while their combination is much more important and valid.

For example, one equation with numerical numbers is provided. One of the methods is a graphical illustration, see Figure 3.1. One simple, but important and interesting case is when all of the coefficients are equal to 1 (i.e., $a = b = k = d = 1$). Equation (3.24) then becomes

$$f(x_1, x_2) = x_1 + x_2 + x_1 x_2 + 1$$

$$
\begin{aligned}
f_0 &= 1   & \hat{f}_0 &= 1 \\
f_1 &= x_1 + 1   & \hat{f}_1 &= x_1 \\
f_2 &= x_2 + 1   & \hat{f}_2 &= x_2 \\
f_{12} &= x_1 + x_2 + x_1 x_2 + 1 & \hat{f}_{12} &= x_1 x_2
\end{aligned}
\tag{3.31}
$$

The zero synergism can be seen in Figures 3.1 and 3.2. There are two axes, x_1 and x_2 ($x_1 = x$ and $x_2 = y$ in the graph).

In Figure 3.1, there are two functions, f_{12} and $\hat{f}_{12}$. For negative or opposite influence, (3.26) yields

$$
\begin{cases} x_1 + x_2 + x_1 x_2 > -1 \\ x_1 x_2 < 0 \end{cases}
\quad \text{or} \quad
\begin{cases} x_1 + x_2 + x_1 x_2 < -1 \\ x_1 x_2 > 0 \end{cases}
\tag{3.32}
$$

This is illustrated in Figure 3.1 by the zoom to the relevant area. The area, in which the functions f_{12}, and $\hat{f}_{12}$ stay on opposite sides of the plane $z = 0$, is exactly the same area in which the synergism is opposite to the initial function.

Figure 3.2 is more complicated, due to the four functions that it includes. There are four functions: f_{12}, $\hat{f}_{12}$, $\hat{f}_1$, and $\hat{f}_2$. For dominant synergism, the equations are

$$
\begin{cases} x_1 + x_2 + x_1 x_2 + 1 > 0 \\ x_1 x_2 > x_1 \\ x_1 x_2 > x_2 \end{cases}
\quad \text{or} \quad
\begin{cases} x_1 + x_2 + x_1 x_2 + 1 < 0 \\ x_1 x_2 < x_1 \\ x_1 x_2 < x_2 \end{cases}
\tag{3.33}
$$

Deeper analyses of the graphs can provide more exact and clear results than mentioned here.

From this example, we move on to a more complicated case.

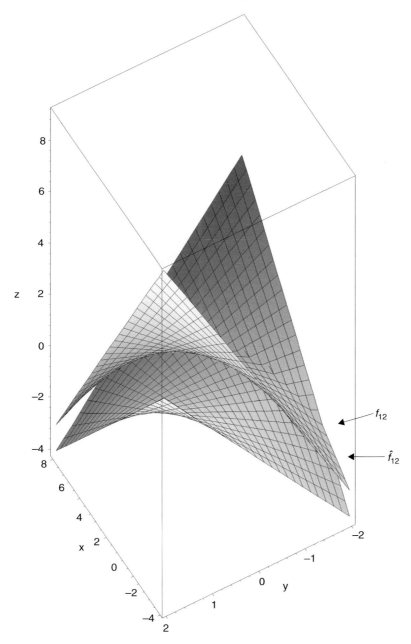

Figure 3.1 Graph for bivariate polynomial for functions f_{12}, $\hat{f}_{12}$, where x-axis and y-axis are x_1 and x_2, and z-axis is the field f. Upper functions f_{12} and lower, $\hat{f}_{12}$. See plate section for color version.

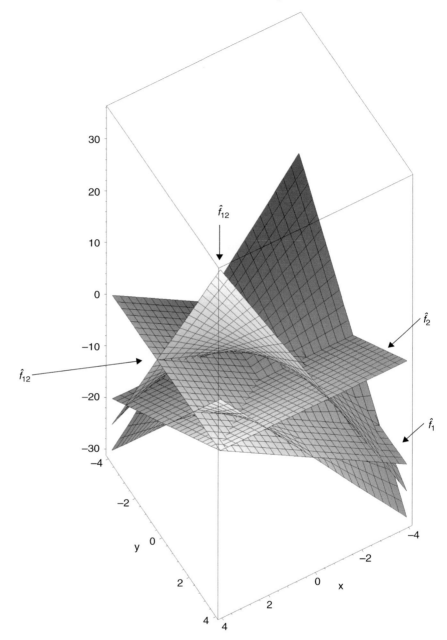

Figure 3.2 Graphs of functions: bivariate polynomial f_{12}, $\hat{f}_{12}$, $\hat{f}_1$, and $\hat{f}_2$. See plate section for color version.

(b) An exponential function $f(x_1, x_2) = e^{ax_1 + bx_2 + d}$

From (3.5), we have

$$f(c_1, c_2) = e^{ac_1x_1 + bc_2x_2 + d}$$

$$f_0 = e^d$$

$$f_1 = e^{ax_1 + d} \qquad (3.34)$$

$$f_2 = e^{bx_2 + d}$$

$$f_{12} = e^{ax_1 + bx_2 + d}$$

and four contributions from (3.6):

$$\hat{f_0} = e^d$$

$$\hat{f_1} = e^{ax_1 + d} - e^d = e^d \left(e^{ax_1} - 1 \right)$$

$$\hat{f_2} = e^{bx_2 + d} - e^d = e^d \left(e^{bx_2} - 1 \right) \qquad (3.35)$$

$$\hat{f_{12}} = e^{ax_1 + bx_2 + d} - e^{ax_1 + d} - e^{bx_2 + d} + e^d = e^d \left(e^{ax_1} - 1 \right) \left(e^{bx_2} - 1 \right)$$

Next, the following three cases are investigated.

(1) The synergism is zero

The expression of $\hat{f_{12}}$ in (3.35) can be zeroed as a result of the trivial case of $x_1 = 0$ or $x_2 = 0$.

(2) The negative interaction

As in the case for polynomial functions, the next inequalities are checked:

$$\begin{cases} f_{12} > 0 \\ \hat{f_{12}} < 0 \end{cases} \quad \text{or} \quad \begin{cases} f_{12} < 0 \\ \hat{f_{12}} > 0 \end{cases} \qquad (3.36)$$

The exponential functions are always positive, so the initial conditions are

$$\begin{cases} f_{12} > 0 \Rightarrow always \\ \hat{f_{12}} < 0 \end{cases} \quad \text{or} \quad \begin{cases} f_{12} < 0 \Rightarrow impossible \\ \hat{f_{12}} > 0 \end{cases} \qquad (3.37)$$

Hence, only the second inequality of the first system has to be checked:

$$e^d \left(e^{ax_1} - 1 \right) \left(e^{bx_2} - 1 \right) < 0 \qquad (3.38)$$

There are two cases in which this inequality (3.38) holds:

$$\begin{cases} e^{ax_1} - 1 < 0 \\ e^{bx_2} - 1 > 0 \end{cases} \quad \text{or} \quad \begin{cases} e^{ax_1} - 1 > 0 \\ e^{bx_2} - 1 < 0 \end{cases} \tag{3.39}$$

If the natural logarithm is taken for both sides of the inequalities, the result is

$$\begin{cases} ax_1 < 0 \\ bx_2 > 0 \end{cases} \quad \text{or} \quad \begin{cases} ax_1 > 0 \\ bx_2 < 0 \end{cases} \tag{3.40}$$

In order to understand what (3.40) means, it is necessary to analyze a more exact case. For example, for the case of

$$\begin{cases} a > 0 \\ b > 0 \end{cases} \tag{3.41}$$

the result means that if the factors are opposing each other, then the synergism works "against" the final effect.

(3) The dominant synergism

$$\begin{cases} f_{12} > 0 \\ \hat{f}_{12} > \hat{f}_1 \\ \hat{f}_{12} > \hat{f}_2 \end{cases} \quad \text{or} \quad \begin{cases} f_{12} < 0 \Rightarrow impossible \\ \hat{f}_{12} < \hat{f}_1 \\ \hat{f}_{12} < \hat{f}_2 \end{cases} \tag{3.42}$$

Substituting the two last inequalities in (3.18) yields

$$\begin{cases} e^d \left(e^{ax_1} - 1\right) \left(e^{bx_2} - 1\right) > e^d \left(e^{ax_1} - 1\right) \\ e^d \left(e^{ax_1} - 1\right) \left(e^{bx_2} - 1\right) > e^d \left(e^{bx_2} - 1\right) \end{cases} \tag{3.43}$$

$$\begin{cases} \left(e^{ax_1} - 1\right) \left(e^{bx_2} - 1\right) > \left(e^{ax_1} - 1\right) \\ \left(e^{ax_1} - 1\right) \left(e^{bx_2} - 1\right) > \left(e^{bx_2} - 1\right) \end{cases} \tag{3.44}$$

Subtracting the left bracket from left and right side:

$$\begin{cases} \left(e^{ax_1} - 1\right) \left(e^{bx_2} - 1\right) - \left(e^{ax_1} - 1\right) > 0 \\ \left(e^{ax_1} - 1\right) \left(e^{bx_2} - 1\right) - \left(e^{bx_2} - 1\right) > 0 \end{cases} \tag{3.45}$$

$$\begin{cases} \left(e^{ax_1} - 1\right) \left(e^{bx_2} - 2\right) > 0 \\ \left(e^{ax_1} - 2\right) \left(e^{bx_2} - 1\right) > 0 \end{cases} \tag{3.46}$$

Four solutions are possible:

$$
\begin{cases}
(e^{ax_1} - 1) > 0 \\
(e^{bx_2} - 2) > 0 \\
(e^{ax_1} - 2) > 0 \\
(e^{bx_2} - 1) > 0
\end{cases}
\quad \text{or} \quad
\begin{cases}
(e^{ax_1} - 1) < 0 \\
(e^{bx_2} - 2) < 0 \\
(e^{ax_1} - 2) > 0 \\
(e^{bx_2} - 1) > 0
\end{cases}
\quad \text{or} \quad
\begin{cases}
(e^{ax_1} - 1) > 0 \\
(e^{bx_2} - 2) > 0 \\
(e^{ax_1} - 2) < 0 \\
(e^{bx_2} - 1) < 0
\end{cases}
$$

$$
\text{or} \quad
\begin{cases}
(e^{ax_1} - 1) < 0 \\
(e^{bx_2} - 2) < 0 \\
(e^{ax_1} - 2) < 0 \\
(e^{bx_2} - 1) < 0
\end{cases}
\tag{3.47}
$$

The second and third systems have opposing conditions, and this yields

$$
\begin{cases}
e^{ax_1} > 2 \\
e^{bx_2} > 2
\end{cases}
\quad \text{or} \quad
\begin{cases}
e^{ax_1} < 1 \\
e^{bx_2} < 1
\end{cases}
\tag{3.48}
$$

Taking the natural logarithm for both sides yields

$$
\begin{cases}
ax_1 > \ln 2 \\
bx_2 > \ln 2
\end{cases}
\quad \text{or} \quad
\begin{cases}
ax_1 < 0 \\
bx_2 < 0
\end{cases}
\tag{3.49}
$$

As in the previous example, we check only one case:

$$
\begin{cases}
a > 0 \\
b > 0
\end{cases}
\tag{3.50}
$$

If the factors are negative, the synergism is always dominant. If the factors do not have the same sign, their common influence is not so important. This means that each factor alone has a greater effect than that of both of them together.

These results show how to obtain specific information about the factors' influences.

(c) A rational function $f(x_1, x_2) = \frac{1}{ax_1 + bx_2 + d}$

For this function, only the first possibility – the zero-synergy case – is checked, because the number of solutions is too large to determine if other than zero-synergy is investigated.

(1) The synergism is zero

The expression of $\hat{f}_{12}$ can be zeroed as a result of the trivial cases: $x_1 = 0$ or $x_2 = 0$, $a = 0$ or $b = 0$, and when the following expression is true:

$$2d + ax_1 + bx_2 = 0$$

As with the first function, for the case of the coefficients for this function equaling 1, we see how the graph of synergism is applied. Then the synergetic functions are

$$
\begin{aligned}
f_0 &= 1 & \hat{f}_0 &= 1 \\
f_1 &= \frac{1}{x_1 + 1} & \hat{f}_1 &= \frac{1}{x_1 + 1} - 1 \\
f_2 &= \frac{1}{x_2 + 1} & \hat{f}_2 &= \frac{1}{x_2 + 1} - 1 \\
f_{12} &= \frac{1}{x_1 + x_2 + 1} & \hat{f}_{12} &= \frac{1}{x_1 + x_2 + 1} - \frac{1}{x_1 + 1} - \frac{1}{x_2 + 1} + 1
\end{aligned}
\tag{3.51}
$$

It is difficult to find all the answers to our questions about the opposite influences of synergism and dominant factors. Figure 3.3 shows the synergic function. As can be seen, the function is discontinuous and has an infinite value at the next points:

$$(x_1 + x_2 + 1)(x_1 + 1)(x_2 + 1) = 0$$

Nevertheless, this is the interpretation of $\hat{f}_{12}$ in (3.51). It fluctuates from $-\infty$ to $+\infty$.

This generates some discussion. Infinity is a term with very distinct, separate meanings that arise in theology, philosophy, mathematics, and everyday life. Popular or colloquial usage of the term often does not accord with its more technical meanings. The word infinity comes from the Latin *infinitus*, "unbounded". In physics, approximations of real numbers are used for continuous measurements and natural numbers are used for discrete measurements (i.e., counting). It is therefore assumed by physicists that no measurable quantity could have an infinite value; for instance, by taking an infinite value in an extended real number system, or by requiring the counting of an infinite number of events. It is, for example, presumed impossible for any body to have an infinite mass or infinite energy. There exists the concept of infinite entities (such as an infinite plane wave), but there are no means to generate such things. Likewise, perpetual motion machines theoretically generate infinite energy by attaining 100% efficiency or greater, and emulate every conceivable open system; the impossible problem follows of knowing that the output is actually infinite when the source or mechanism exceeds any known and understood system. This point of view does not mean that infinity cannot be used in physics.

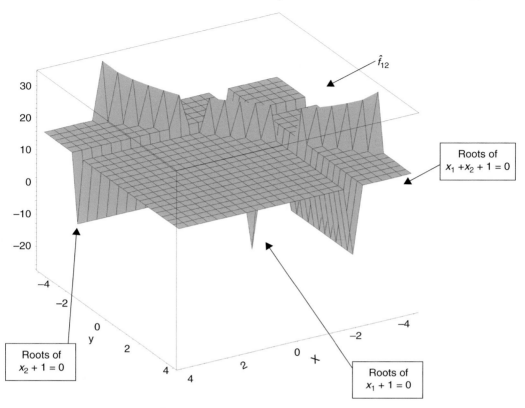

Figure 3.3 Graph of $\hat{f}_{12}$ for rational function. x_1, x_2 are represented by x, y.

To sum up, in most of the cases in which the synergy function starts to fluctuate from $-\infty$ to $+\infty$, many other factors try to offset this. An example of this in the atmosphere can be a developing cyclone that typically has a preceding warm front and a faster-moving cold front. As the storm intensifies, the cold front rotates around the storm and catches the warm front. From this point the cyclone is extinguished.

All of these simple functions and research on synergisms provide us with a basis for working with more complicated functions and cases in the following chapters.

4

Factor Separation Methodology and paleoclimates

A. Berger, M. Claussen, and Q. Yin

The relative contribution of the individual forcings, the feedbacks and the synergisms can be quantified by using different techniques, such as the Alpert–Stein Factor Separation Methodology (here abbreviated FS; Stein and Alpert, 1993) or other feedback analyses. In an attempt to better understand the role of the temperature–albedo feedback of the greenhouse gases (water vapour and CO_2), and of the insolation at the Last Glacial Maximum (LGM), several sensitivity experiments have been made with a radiative convective model (Berger *et al.*, 1993) and results discussed using both the classical feedback analysis and FS methodology. The LGM cooling is simulated to be $4.5\,°C$, of which $3\,°C$ is in response to the insolation–albedo forcing and $1.5\,°C$ is from the CO_2 forcing. In these experiments, the water vapour feedback (WVF) is included, but the synergisms appear to be very small. The direct influence of the insolation–albedo forcing is a cooling of $1.8\,°C$, on top of which the WVF adds $1.2\,°C$. The remaining $1.5\,°C$ is due to the CO_2 forcing, of which $0.9\,°C$ comes from its direct influence and $0.6\,°C$ is due to the WVF feedback.

The FS methodology and a generalisation of the classical linear feedback analysis technique are also used to identify the individual contributions of climatic factors and of their synergism to the Holocene climate change signal. From the Ganopolski *et al.* (1998) experiments it can be shown that for the temperature differences between 6 ka BP (6000 years before present) and the present in boreal latitudes, the synergism due to changes in the vegetation cover, sea-surface temperatures and the sea-ice extent plays a more important role than the pure contributions of the ocean and of the vegetation themselves. Contrary to temperature, precipitation changes in summer over the Northern Hemisphere continents – North Africa in particular – are mainly related to the pure contribution of vegetation changes. Advantages and disadvantages of the different methods in such kinds of analyses are discussed.

Factor Separation in the Atmosphere: Applications and Future Prospects, ed. Pinhas Alpert and Tatiana Sholokhman. Published by Cambridge University Press. © Cambridge University Press 2011.

4.1 Introduction

Comprehensive climate system models numerically describe the evolution of the dynamic and thermodynamic states of the atmosphere, the ocean, the terrestrial and marine biosphere, and – some – the ice sheets, in response to a given forcing, from a set of equations based on physical and chemical laws. Because of the many variables, equations and non-linear interactions that control the behaviour of the climate system, it is very difficult to trace in the model what the direct perturbations induced by the forcings become with time. One way to investigate the problem is to perform sensitivity experiments. Their purpose is to isolate the effects of different factors (either a forcing such as the incoming solar radiation or a feedback such as water vapour) on certain climate variables (such as temperature or ice volume). The most common method of evaluating the contribution of a specific factor is by analysing the difference between the climate fields of the control run, where the influence of this factor is switched on, and of a simulation, where the influence of this factor is switched off. However, it has been shown by Stein and Alpert (1993) that the difference map of the two simulations does not have a simple meaning, especially when more than two factors are considered.

For example, when investigating the effects of the albedo and water vapour feedbacks in a double-carbon dioxide (CO_2) or astronomical experiment, three simulations are generally performed: the control simulation (CON), which allows the two feedbacks to amplify the initial perturbation induced by doubling the CO_2 concentration or by changing the seasonal and latitudinal distribution of solar radiation; the constant albedo simulation (NA); and the constant water vapour simulation (NWV). Actually, the difference between the simulated fields of CON and NA shows not only the direct effect of the albedo feedback, but also the indirect effect of the water vapour feedback resulting from the temperature change that appears when the albedo is allowed to vary. The same holds for the difference between CON and NWV, which shows the direct effect of the water vapour and the indirect effect related to the albedo change when the water vapour is allowed to vary. If these interactions between the two feedback loops are not identified, the differences (CON–NA) and (CON–NWV) cannot be simply interpreted, contrary to what is commonly attempted.

Such analysis of the relative importance of the feedback mechanisms involving CO_2, water vapour, and albedo has been made, in particular from transient experiments with the LLN 2-D model (Gallée *et al.*, 1991). The response of the model to the astronomical and the CO_2 forcings showed that, at the Last Glacial Maximum (LGM), the lower CO_2 atmospheric concentration might have been responsible for roughly 50% of the cooling and for 30% of the ice-volume change (Gallée *et al.*,

1992). In an attempt to better understand the role of the temperature–albedo feed-back, of the greenhouse gases (water vapour and CO_2) and of the insolation at this LGM, several additional sensitivity experiments have been made with a radiative convective model (Berger *et al.*, 1993) and results discussed using both the classical feedback analysis and the FS methodology by Stein and Alpert (1993) (Section 4.2).

Another illustration to quantify the feedbacks in a climate model will be based on the analysis of the relative role played by the ocean and the vegetation at the LGM from the simulation made by Ganopolski *et al.* (1998). An extended feedback tech-nique by Claussen (2001) and additive amplifying factors introduced by Kubatzki *et al.* (2000) will be compared to the Alpert–Stein methodology in Section 4.3.

4.2 Insolation, albedo, and greenhouse gases at the LGM

4.2.1 Forcing, feedbacks, and response

Let N be the net energy flux at a given level (or any climatic predicted field) and T_s, the surface temperature. N is a function of T_s, and of external, E, and internal, I, parameters. E is a vector of quantities – called forcings – whose changes can lead to a change in climate, but which are independent of climate. I is a vector of quantities that can change as the climate changes and feed back to modify the initial climatic perturbation (Schlesinger, 1989). E includes, for example, the solar constant, the volcanic dust, the insolation and, in our experiments, the CO_2 atmospheric concentration (although it is actually a feedback that changes as a result of climatic change). I contains the variables of the climate system, except T_s (e.g., the albedo α, and the water vapor e). If T_s is the only dependent variable in a model, as it is in energy balance models, $I = I(T_s)$.

A small change in N, ΔN, is generally expressed as:

$$\Delta N = \sum_i \frac{\partial N}{\partial E_i} \Delta E_i + \left(\frac{\partial N}{\partial T_s} + \sum_j \frac{\partial N}{\partial I_j} \frac{dI_j}{dT_s} \right) \Delta T_s \qquad (4.1)$$

or

$$\Delta N = -\Delta Q + \lambda \Delta T_s \qquad (4.2)$$

which defines by direct identification with (4.1):
the direct radiative forcing ΔQ,

$$\Delta Q = -\sum_i \frac{\partial N}{\partial E_i} \Delta E_i$$

the feedback parameter λ,

$$\lambda = \frac{dN}{dT_s} = \frac{\partial N}{\partial T_s} + \sum_j \frac{\partial N}{\partial I_j} \frac{dI_j}{dT_s}$$

After the equilibrium is reached in response to the external forcing ΔQ, $\Delta N = 0$ and we have

$$\Delta T_s = \frac{\Delta Q}{\lambda} = G \Delta Q = \frac{\Delta Q}{\lambda_0 - F} = \frac{G_0}{1 - f} \Delta Q$$

where:

$G = 1/\lambda$ is the gain (output/input) of the climate system

$\lambda_0 \Delta T_s$ is the change in N due to the change of T_s alone; $\lambda_0 = \frac{\partial N}{\partial T_s}$

$F \Delta T_s$ is the change in N due to the change in the internal quantities through their dependence on T_s:

$$F = -\sum_j \frac{\partial N}{\partial I_j} \frac{dI_j}{dT_s}$$

$G_0 = 1/\lambda_0$ is the gain (output/input) of the climate system when the feedback factor $f = G_0 F = 0$. Thus G_0 is the zero-feedback gain and $(\Delta T_s)_0 = G_0 \Delta Q$ is the temperature change without the feedbacks.

R and ΔT_s are the gain factor and temperature change with the feedbacks (the response of the climate system to the forcings):

$$\frac{\Delta T_s}{(\Delta T_s)_0} = \frac{1}{1 - f} = R = \frac{G}{G_0}$$

Classically, to understand the sensitivity of the climate system to some external and/or internal parameters, we change one by one the values of the factors we need to quantify the impact. In addition to the simulation of the present-day and LGM climates in response to the insolation and CO_2 forcings, experiments will therefore be performed changing them successively, allowing the water vapour to vary or not.

4.2.2 Application to the LGM cooling

A radiative–convective model (RCM) has been used to analyse the individual role played by the insolation, albedo and greenhouse gases (CO_2 and water vapour) at the LGM. The main reason for using a RCM was related to the number of experiments to be done to isolate the individual contribution of the main processes involved

Table 4.1 *Sensitivity experiments of a radiative–convective model to CO_2, water vapour and surface albedo (Berger* et al., *1993)*

Experiment	CO_2 (ppmv)	Albedo α_s (%)	Water vapour (g cm^{-2})	V/F	T_s (K)	ΔQ (W m^{-2})
1 (present day)	330	16.24	2.289	V	288.32	0
2	194	16.24	2.289	F	287.38	−2.91
3	330	19.25	2.289	F	286.52	−5.62
4	194	19.25	2.289	F	285.58	−8.53
5	330	16.24	1.697	F	286.58	0
6	194	16.24	2.07	V	286.75	−2.91
7	330	19.25	1.88	V	285.33	−5.62
8 (LGM)	194	19.25	1.697	V	283.81	−8.53
9	330	19.25	1.697	F	284.76	−5.62
10	194	16.24	1.697	F	285.63	−2.91

V/F water vapour feedback switched on and off, i.e., water vapour allowed to vary (V) or
    kept fixed (F).
ΔQ initial radiative perturbation at the tropopause after a modification of the CO_2
    concentration and of the surface albedo (reference value is present day).
T_s annually and hemispherically averaged temperature at the surface.

and to the potentiality of a RCM for reaching such an objective. At the LGM, in addition to the insolation and CO_2-concentration change, changes in the surface albedo and in the vertical distribution of water vapour were inserted one by one or in combination into the RCM. As a consequence, the only feedback mechanism in this RCM is related to water vapour. The main contribution of each factor to the LGM cooling has been calculated using the classical and the Alpert–Stein methodology of feedback analysis, and the results are summarised in the next sections.

RCM Experiment 1 (Table 4.1) is related to the present-day climate and leads to a mean temperature of 288.32 K. In Experiment 3, only the surface albedo is changed to its LGM value, the CO_2 and water vapour concentration being kept constant to their present-day value. In such a case, if we accept the Milankovitch hypothesis according to which the insolation variations modify the snow and ice covers in the northern hemisphere high latitudes, changing only α_s in the RCM is equivalent to looking for the 'natural' response of the climatic system to the insolation variations. The simulated cooling of 1.8 °C is important and corresponds to a direct radiative forcing of −5.6 W m^{-2} (i.e., the net radiative budget at the tropopause has decreased by 5.6 W m^{-2}). If we allow the water vapour feedback to amplify this perturbation (Experiment 7), the cooling amounts to approximately

3 °C, which corresponds to a gain factor of 1.66. The smaller amount of water vapour in this cooler atmosphere leads indeed to a weakening of the greenhouse effect. It combines with the larger planetary albedo induced by the increase of the surface albedo to produce a total cooling 66% larger than without the water vapour feedback. As a consequence, the water vapour induces a positive feedback mainly through its impact on the greenhouse effect and, by a much smaller amount, through an increase in the planetary albedo related to a smaller absorption of the solar radiation.

If we decrease only the CO_2 concentration from the assumed present-day value (330 ppmv) to the LGM value (194 ppmv), the climatic response is a cooling of 0.94 °C or of 1.57 °C, if the experiments are made without (Experiment 2) or with (Experiment 6) the water vapour feedback, respectively. This corresponds to a gain factor of 1.67, very similar to the value obtained from Experiments 3 and 7 for the albedo increase. This might be a coincidence, and must be tested using more sophisticated models where geographical contrasts do exist. The change of the surface albedo at the LGM (Experiments 3 and 7) is indeed largely related to the change of the surface cover in middle and high latitudes, whereas the CO_2 concentration change (Experiments 2 and 6) is a worldwide phenomenon.

Experiments 4 and 8 combine the changes in the CO_2 concentration and in the surface albedo. When the water vapour feedback is switched off (Experiment 4), the cumulative effect of these forcings leads to a total cooling that is the sum of the individual coolings: $-2.74 = -1.8 - 0.94$. This is expected through a linearisation of the problem in the absence of the main non-linearity related to the water vapour feedback. Allowing this water vapour feedback to amplify the direct initial perturbation leads to a total cooling of 4.51 °C (Experiment 8), very close to the cooling at the LGM estimated from the proxy records and calculated by general circulation models. Even in this case, the total response is more or less the sum of the individual responses (Experiments 7 and 6): $-4.51 \cong -2.99 - 1.57$. This is because the perturbations in amplitude are actually small as compared to the fundamental state (a few degrees against 288 K).

The feedback factor, $f = 1 - 1/R$, for the water vapour estimated from Experiments 6, 7 and 8 appears to be roughly 0.4. λ is equal to 3.1 in Experiments 2, 3 and 4 and to roughly 1.9 in Experiments 6, 7 and 8; this leads to a positive feedback of the water vapour equal to 1.2, a value close to the value 1.4 obtained in the greenhouse theory (e.g., Berger and Tricot, 1992). It happens therefore that, in our RCM, the water vapour feedback intensity is almost independent of the specific character of the prescribed changes. This again might be a characteristic of this model and must be tested with more sophisticated GCMs.

A. Berger, M. Claussen, and Q. Yin

Table 4.2 *Summary of the contributions of CO_2 forcings and of the water vapour feedback (WVF) and astronomical-insolation-albedo (AA) to the 4.5 °C cooling at the LGM*

	ΔT (°C)		ΔT (%)	
WVF + AA	1.8	AA	40	AA
	1.2	WVF	27	WVF
CO_2 + WVF	0.6	WVF	13	
	0.9	CO_2	20	CO_2

(ΔT range marked 4.5 at top, 1.5 at middle)

All these sensitivity tests using the value of the parameters at the LGM lead to the following conclusions for the northern hemisphere (Table 4.2):

1. The cooling using only the astronomical-insolation-albedo (AA) forcing – but allowing for the water vapour feedback – amounts to 3 °C, 67% of the 4.5 °C cooling resulting from both the AA and CO_2 forcings.
2. If the CO_2 concentration is kept fixed to its interglacial level (330 ppmv), the AA forcing without the water vapour feedback (WVF) explains 60% of the 3 °C cooling, the water vapour feedback being therefore responsible for 40% of the total cooling in this astronomical-insolation–albedo alone experiment.
3. In the (AA + CO_2) experiment, the direct effect of these forcings (i.e., without the WVF) explains 60% (2.7 °C) of the 4.5 °C cooling. As in 2, the WVF is therefore responsible for 40% of the cooling. If we discriminate between the AA forcing and the CO_2 contribution, we can see that without the WVF, AA explains 40 of the 60% of the total cooling. The contributions to the WVF (40% of the total cooling) are 27% for AA and 13% for CO_2.
4. In the (AA + CO_2) cooling, it may also be interesting to analyse the proportion that is due to water vapour for each sub-experiment. Considering only the AA forcing, the WVF is responsible for 27 of the 67% of the global cooling, the remaining 40% is the direct effect of AA. In the CO_2 alone experiment, the WVF explains 13 of the 33% of the global cooling, the direct effect of CO_2 accounting for the remaining 20%.

But in this theory, the formula holds only for small perturbations around a reference state and does not consider the synergism between the variables. To account for these higher order interactions, the Alpert–Stein method (Stein and Alpert, 1993) must be used.

4.2.3 The Alpert–Stein Factor Separation Methodology

The Alpert–Stein methodology aims to identify the contribution to a predicted field, y, of the processes (factors) involved in a climate model as well as their synergistic effects. When applied to two factors (astronomical-insolation-albedo and CO_2 in our previous example), four experiments are required. In the following, the predicted field will be given the value y_{11} when all factors are acting, 'switched on', and a value y_{00} when they are all 'switched off'. In the two other experiments, y_{10} and y_{01}, only one factor is changed at a time. According to Stein and Alpert (1993), y can be decomposed through a series expansion as follows:

$$y_{11} = y_{00} + (y_{10} - y_{00}) + (y_{01} - y_{00}) + (y_{11} - y_{10} - y_{01} + y_{00})$$

$$y_{11} = y_{00} + \widehat{y}_{10} + \widehat{y}_{01} + \widehat{y}_{11}$$

where $\widehat{y}_{10} = y_{10} - y_{00}$ is the pure contribution of the first factor

$\widehat{y}_{01} = y_{01} - y_{00}$ is the pure contribution of the second factor

$\widehat{y}_{11} = y_{11} - (y_{10} + y_{01}) + y_{00}$ is the contribution of the synergism between

the two.

To test the sensitivity of a model to a particular factor, one usually compares the results of an experiment where its effect is not active (for example, y_{10} if the second factor is switched off) to the control experiment, y_{11}. This difference is actually equal to

$$y_{11} - y_{10} = y_{11} - y_{00} - \widehat{y}_{10} = \widehat{y}_{01} + \widehat{y}_{11}$$

This clearly indicates that the traditional method does not provide the expected pure contribution of the process analysed, but this contribution plus the contribution of its interactions with the others. This is usually implicitly accounted for in the line of argument developed in sensitivity analysis, but the FS methodology provides an easy way to objectively separate the pure contribution of one factor from its synergism with the others.

This technique will be applied to our preceding case to determine the pure contributions of the astronomical-insolation-albedo, of CO_2 and of their synergism through the water vapour feedback when the climate changes from the present-day to the LGM or vice-versa.

Table 4.3 *Carbon dioxide and surface albedo: no water vapour feedback; LGM reference*

	Experiment	CO_2	α_s	WV	T_s	ΔT_s
				Water vapour fixed present-day value; PR = LGM + $\sum$		
y_{11}	1	330	16	2.3	288.3	2.74
y_{00}	4	194	19	2.3	285.6	—
y_{10}	3	330	19	2.3	286.5	0.94
y_{01}	2	194	16	2.3	287.4	1.80

Units and symbols as in Table 4.1. ΔT_s is the deviation from the y_{00} value.

Referring to Table 4.1, we will first consider the experiments where the water vapour feedback is kept fixed to its present-day value (Experiments 1 to 4). In such a case (Table 4.3), there is no relation at all between CO_2 and the albedo (no synergism). Starting from the LGM, y_{00} (Experiment 4), the present-day climate, designated by y_{11} (indicating that the two variables are 'on', Experiment 1), is given by

$$y_{11} = y_{00} + \widehat{y}_{10} + \widehat{y}_{01} + \widehat{y}_{11}$$

where $\widehat{y}_{10}$ is obtained through Experiment 3 (CO_2 in) and is equal to: $y_{10} - y_{00} = 0.94°$ C. $\widehat{y}_{01}$ is obtained through Experiment 2 (α_s in) and is equal to: $y_{01} - y_{00} = 1.8°C$ and therefore $\widehat{y}_{11} = 2.74° C - 0.94°C - 1.8°C = 0°C$ as expected.

Taking into account the water vapour feedback (Table 4.4a, Experiments 1, 6, 7, 8) leads to

$$\widehat{y}_{10} \quad \text{(Experiment 7)} = 1.52°C$$
$$\widehat{y}_{01} \quad \text{(Experiment 6)} = 2.94°C$$
$$\text{and} \quad \widehat{y}_{11} = 4.51 - 1.52 - 2.94 = 0.05°C$$

This clearly shows that the warming from the LGM to the present day is due to CO_2 for 33.7%, to the insolation–albedo for 65.2% and to their synergism for 1.1%. This synergism is due to the water vapour, which is the only common factor between the two variables. A similar result is obtained when the cooling at the LGM is calculated starting from the present day (Table 4.4b).

Table 4.4a *Carbon dioxide and surface albedo: with water vapour feedback; LGM reference*

	Experiment	CO_2	α_s	WV	T_s	ΔT_s
		Water vapour variable; $PR = LGM + \sum$				
y_{11}	1	330	16	2.3	288.3	4.51
y_{00}	8	194	19	1.7	283.8	—
y_{10}	7	330	19	1.9	285.3	1.52
y_{01}	6	194	16	2.1	286.8	2.94

Table 4.4b *Carbon dioxide and surface albedo: with water vapour feedback; present-day reference*

	Experiment	CO_2	α_s	WV	T_s	ΔT_s
		Water vapour variable; $LGM = PR + \sum$				
$\underline{y}_{11}$	8	194	19	1.7	283.8	−4.51
$\underline{y}_{00}$	1	330	16	2.3	288.3	—
$\underline{y}_{10}$	6	194	16	2.1	286.8	−1.57
$\underline{y}_{01}$	7	330	19	1.9	285.3	−2.99

Units and symbols as in Table 4.1.

If the three variables, CO_2, insolation–albedo and water vapour, are used, the 4.5 °C warming from the LGM to the present day is estimated to comprise:

from CO_2, 0.95 °C
from insolation–albedo, 1.84 °C
from water vapour, 1.77 °C

The synergisms between these two or three factors are weak (less than 0.1 °C), but with three factors, they slightly reduce the excessive warming accounted for by the individual factors. In both cases, this is why the results obtained with the Alpert–Stein methodology are very similar to those obtained with the classical method. This is not the case when the synergisms are strong, as demonstrated in the next section.

4.3 Vegetation and ocean during the mid-Holocene

4.3.1 Introduction

Feedback analysis techniques can also be applied to test the role played by the different components of the climate system, such as the ocean and the continental biosphere. Different kinds of climate models have attempted to simulate such an impact of vegetation changes on climate, for example over the last glacial–interglacial cycle (Berger, 2001), during the Eemian (Kubatzki *et al.*, 2000; Crucifix and Loutre, 2002), at the Last Glacial Maximum (Kubatzki and Claussen, 1998; Jahn *et al.*, 2005) or during the Holocene (Claussen and Gayler, 1997; Ganopolski *et al.*, 1998; Crucifix *et al.*, 2002; Otto *et al.*, 2009a). Vegetation changes are important for climate because the surface albedo and roughness, and thus the energy budget, are not only related to the snow and ice fields, but also to the nature of vegetation that covers the land surfaces (Charney *et al.*, 1977; Claussen, 2004). All these modeling efforts show that changes in vegetation cover modify and amplify the response of the climate system to the astronomical forcing, both directly primarily through the changes in surface albedo, and indirectly through changes in oceanic temperature, sea-ice cover and oceanic circulation.

The FS methodology and an extended feedback analysis will be used to objectively quantify the individual contributions of ocean, vegetation, and their synergism to the temperature and precipitation changes of the mid-Holocene as found in Ganopolski *et al.* (1998). For the mid-Holocene, Ganopolski *et al.* have made four experiments with different model configurations, from the atmosphere-only model to the coupled atmosphere–ocean–terrestrial vegetation model, including coupled atmosphere–ocean and atmosphere–vegetation models where, respectively, vegetation and ocean were fixed to their simulated present-day state. In the following, A will denote the value of a climate variable (temperature or precipitation) calculated by the atmosphere-only model, AO by the atmosphere–ocean model, AV by the atmosphere–vegetation model and AOV by the coupled atmosphere–ocean–vegetation model. To reach their goal, Ganopolski *et al.* have analysed the differences from the present-day control run, R, of their four mid-Holocene simulations. Along with these notations, the FS methodology leads to

$$AOV - R = A - R + O + V + OV \qquad (4.3)$$

where

$$O = AO - A \qquad (4.4)$$

gives the pure contribution of the ocean

$$V = AV - A \qquad (4.5)$$

gives the pure contribution of vegetation

$$OV = AOV - AO - AV + A \qquad (4.6)$$

gives the contribution of the synergism between the ocean and vegetation, and

$$AOV - OV \qquad (4.7)$$

gives the reaction of the fully coupled system without taking into account the synergism.

Kubatzki *et al.* (2000) expanded this FS methodology in a difference form because, in the original version of this method, the contribution of a given factor is estimated from experiments where this factor is strictly switched on or off and not necessarily given a reference value (Alpert *et al.*, 1996). They demonstrated that, using the deviations with respect to the control/reference run, R, instead of using the absolute values of the variables, does not affect any of the Stein–Alpert equations. In the following we will continue along the same line and replace the values of A, AO, AV and AOV by their deviations from the reference, i.e., by $A - R$, $AO - R$, $AV - R$ and $AOV - R$. This leads to the definition of the Kubatzki *et al.* additive amplification factors:

$$f_a = a/(A - R) \quad \text{with} \quad a = AOV - R, O, V, OV \qquad (4.8a)$$

and equation (4.3) normalised by $(A - R)$ becomes

$$f_{AOV} = 1 + f_O + f_V + f_{OV} \qquad (4.8b)$$

The additive amplification factors are therefore giving the pure contributions of the ocean, the vegetation, their synergism or all the factors relative to the change of climate from 6 ka BP to the present day induced by the atmosphere only. A positive amplification factor thus means an amplification of the pure response of the atmospheric model (i.e., without feedbacks from the other sub-systems) due to the reaction of the ocean, vegetation, their synergism or the combination of all components; a negative factor means a weakening. Values lower than one show that the amplification/weakening is smaller than the effect of the atmosphere alone and thus cannot compensate it; values larger than unity show that it is stronger. This method was discussed in Kubatzki *et al.* (2000).

4.3.2 Extended feedback analysis

On the other hand, in order to compare more adequately the FS methodology with the conventional feedback analysis, Claussen (2001) extended such analysis to include the synergisms. S, the response of the climate system to an ensemble of external forcings, E, can indeed be written in terms of a gain G: $S = GE$. Without any feedback, the response of the system would be $S_0 = G_0 E$. With feedbacks, however, one has to assume that the response S of the full system is modified by some internal or feedback processes H_i, which are triggered by S. The outputs $(\sum_i H_i)S$ of these internal processes feed into the system such that

$$S = G_0 E + G_0 \left(\sum_i H_i \right) S \qquad (4.9)$$

The factor $G_0 H_i$ is called feedback f_i. Hence $G = G_0/(1 - \sum f_i)$. Feedbacks f_i can be evaluated by setting up a number of sensitivity experiments to obtain

$$S_i = G_0 E + f_i S_i \qquad (4.10)$$

where S_i is the output signal of the system in the case that only one internal process H_i is operating. Peixoto and Oort (1992) mention, however, that this analysis is based on the assumption that there exists no synergism, i.e., interaction among feedbacks.

To include these interactions, Claussen (2001) defined a multidimensional transfer function $\widehat{G}$, which not only includes feedbacks, but also synergisms between feedbacks. Formally, one may write: $\widehat{S} = \widehat{G}E$, where $\widehat{S}$ is the response of the full system, and

$$\widehat{S} = G_0 E + \left(\sum_i f_i \right) \widehat{S} + \left(\sum_i \sum_j f_{ij} \right) \widehat{S} + \left(\sum_i \sum_j \sum_k f_{ijk} \right) \widehat{S} + \cdots \quad (4.11a)$$

where f_{ij} indicate the synergism between two processes H_i and H_j (with $i \neq j$), and f_{ijk} between H_i, H_j, and H_k (with $i \neq j \neq k$). As for feedbacks, we can differentiate between positive synergisms, $f_{ij} > 0$, i.e., synergisms that strengthen the full signal such that $\widehat{S}$, and negative ones, $f_{ij} < 0$, which weaken the output $\widehat{S}$ in comparison with S. The feedbacks and synergisms that will be defined here differ from those introduced originally by Claussen (2001) by the normalising factor. In Claussen (2001), f_1 was defined by $\frac{O}{AO}$, whereas it will be given here by $\frac{O}{AO-R}$. The same holds for f_2 and f_{12}, so that the conclusions will be slightly different, as the denominator is no longer an absolute value but a deviation from the present-day reference.

To compare the FS methodology with the extended feedback analysis, we rewrite Eq. (4.11a) for the two-dimensional case, taking into account that it deals with the deviations from the reference run, as

$$AOV - R = A - R + f_1(AOV - R) + f_2(AOV - R) + f_{12}(AOV - R)$$
(4.11b)

In this notation, given (4.10) where in a two-factor analysis (ocean and vegetation) $G_0 E$ can be identified with the atmosphere-only model with no feedback and S_i with the atmosphere–ocean (AO) and atmosphere–vegetation (AV) models, we obtain

$$f_1 = (AO - A)/(AO - R) \qquad f_2 = (AV - A)/(AV - R)$$
(4.11c)

which means from (4.3) and (4.4) that the pure contribution of the ocean is given by

$$O = f_1(AO - R)$$
(4.12a)

and the pure contribution of vegetation by

$$V = f_2(AV - R)$$
(4.12b)

f_1 and f_2 are therefore the amplification factors of the atmosphere–ocean and of the atmosphere–vegetation models (not of the atmosphere only, which is the case for f_O and f_V) to provide the pure contribution of, respectively, the ocean and vegetation.

Using (4.11b),

$$\frac{A - R}{AOV - R} = 1 - (f_1 + f_2 + f_{12})$$

and comparing to (4.8b)

$$\frac{AOV - R}{A - R} = 1 + f_O + f_V + f_{OV}$$

we obtain

$$1 + f_O + f_V + f_{OV} = \frac{1}{1 - f_1 - f_2 - f_{12}}$$
(4.13)

From (4.8a) and (4.12), we obtain

$$(1 + f_O) = \frac{1}{(1 - f_1)} \qquad (1 + f_V) = \frac{1}{(1 - f_2)}$$
(4.14)

showing that for small feedbacks $f_a \simeq f_i$ for $a = O, V$ and $i = 1, 2$.

From equations (4.12a,b) it becomes obvious that feedbacks f_i indicate the ratio of the pure contribution of the ocean $(AO - A)$ and of the vegetation $(AV - A)$ to the deviation from the reference of the full signal $(AO - R$ or $AV - R)$ in an interactive atmosphere–ocean (AO) or atmosphere–vegetation (AV) simulation, respectively. The amplification factors f_a indicate the amplification of the initial difference $A - R$ to obtain the pure contribution (O or V). Hence feedbacks f_i and amplification factors f_a denote pure contributions of internal processes in a similar way

$$O = f_O(A - R) = f_1(AO - R)$$
$$V = f_V(A - R) = f_2(AV - R)$$

However, because of the different normalisation ($AO - R$ and $AV - R$ in the case of f_i, and $A - R$ in the case of f_a) interpretation slightly differs, as will be discussed in Section 4.4. Moreover, the sign of f_a and f_i will be the same if it is also true for $A - R$ and $AO - R$ or $AV - R$.

Finally, the synergism between the ocean and vegetation, OV, can be written in the three configurations. From the Alpert–Stein methodology, we have

$$OV = (AOV - R) - (A - R) - O - V \tag{4.15a}$$

Using (4.8a), OV can be expressed in terms of the additive amplifying factors

$$OV = (A - R)(f_{AOV} - 1 - f_O - f_V) \tag{4.15b}$$

Replacing AOV in (4.15a) by its value from (4.11b) allows to link the OV of Alpert–Stein to the Claussen feedbacks f_i

$$OV = f_{12}(AOV - R) + f_1(AOV - AO) + f_2(AOV - AV) \tag{4.15c}$$

Using the different relationships defining and relating the feedbacks f_i and the amplifying factors f_a, we obtain successively:
from Alpert–Stein

$$(A - R)f_{OV} = (AOV - R) - (A - R) - O - V \tag{4.16a}$$

from (4.8a) using the factors f_a

$$f_{OV} = f_{AOV} - 1 - f_O - f_V \tag{4.16b}$$

or

$$f_O + f_V + f_{OV} = \frac{AOV - A}{A - R} \tag{4.16c}$$

from (4.13) and (4.14) with the feedbacks f_i

$$fov = \frac{f_1 + f_2 + f_3}{1 - (f_1 + f_2 + f_3)} - \frac{f_1}{1 - f_1} - \frac{f_2}{1 - f_2} \qquad (4.16d)$$

In a similar way, starting from (4.11b), f_{12} can be expressed in three different ways

$$f_{12} = \frac{AOV - A}{AOV - R} - \frac{AO - A}{AO - R} - \frac{AV - A}{AV - R} \qquad (4.17a)$$

$$f_{12} = \frac{fo + fv + fov}{1 + (fo + fv + fov)} - \frac{fo}{1 + fo} - \frac{fv}{1 + fv} \qquad (4.17b)$$

$$f_{12} = \frac{AOV - A}{AOV - R} - f_1 - f_2 \qquad (4.17c)$$

$$\text{or } f_1 + f_2 + f_{12} = \frac{AOV - A}{AOV - R}$$

Comparing f_{12} to fov is however less straightforward. We have indeed

$$OV = fov(A - R)$$

which has to be compared to (4.15c) to link f_{12} to fov:

$$fov(A - R) = f_{12}(AOV - R) + f_1(AOV - AO) + f_2(AOV - AV) \qquad (4.18)$$

Hence, factor fov indicates the amplification of the initial signal difference $A - R$ to obtain the strength of the contribution OV owing to synergisms. From (4.17c), f_{12}, on the other hand, indicates the additive contribution, $f_{12}(AOV - R)$, of a synergism to the difference between the full signal AOV and the initial signal A to obtain the contributions $f_i(AOV - R)$ related to the feedbacks f_1 and f_2. The numerical value of f_{12} is given by the difference from the sum $f_1 + f_2$ of the relative contribution of the deviation from the atmosphere only (A, no feedback, no synergism) of the full experiment AOV, to the deviation of AOV from the reference, R. Actually, the sum $fo + fv + fov$ is a measure of the difference between the results of the full experiment AOV and of the atmosphere-only A, relative to $A - R$ (see 4.16c), although the sum $f_1 + f_2 + f_{12}$ is a measure of the same difference, but relative to $AOV - R$ (see 4.17c). The conclusions that can be drawn from the additive amplification factors f_a (see 4.8a) or the feedbacks f_i (see 4.11c) and their related synergisms fov and f_{12} are complementary, using either the atmosphere only ($A - R$) or any coupled climate system (AO, AV, AOV) as a reference.

4.3.3 Application to the feedbacks of the ocean and vegetation

In the atmosphere-only model, there is no component with a long response time, so we can assume that it will respond linearly to the astronomical forcing shown in

Figure 4.1. The difference in the seasonal and latitudinal distribution of the energy received from the Sun between the peak of the Holocene (6 ka BP) and the present day originates from precession. Today, northern hemisphere (NH) winter occurs at perihelion (the longitude of the perihelion is 102°), whereas at 6 ka BP it was the fall equinox that occurred at the minimum Earth–Sun distance (longitude of the perihelion was close to 0°). As seen on Figure 4.1, the Earth received more energy at 6 ka BP than presently, from the North Pole in July, to the equator in August and finally to the South Pole in October (with a maximum of respectively 8, 7 and 5%). This is balanced by a deficit centred just south of the equator in February and extending symmetrically to the north and to the south and from the winter solstice to the spring equinox. The Earth therefore received more energy during NH summer and less during NH winter, although this is less sharp than in the comparison between times when NH summer occurs either at perihelion or at aphelion. At the summer solstice, the NH received about $20\,\mathrm{W\,m^{-2}}$ more at 6 ka BP than now and the southern hemisphere (SH) only about $2\,\mathrm{W\,m^{-2}}$ more. This is the origin of the warming of $1\,°\mathrm{C}$ and of $0.2\,°\mathrm{C}$ of the NH and the SH, respectively, in the simulation using only the atmosphere (Table 4.5). At the winter solstice, the NH and SH received about $10\,\mathrm{W\,m^{-2}}$ less at 6 ka BP than now, a deficit reflected in the NH and SH cooling of 0.5 and $0.3\,°\mathrm{C}$, respectively.

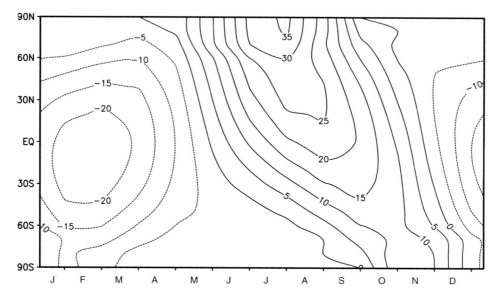

Figure 4.1 Latitudinal and seasonal distribution of the difference in the energy received from the Sun between 6 ka BP and the present day (in $\mathrm{W\,m^{-2}}$; Berger, 1978)

Table 4.5 *Surface air temperature in the atmosphere–ocean–vegetation models at the peak of the Holocene (6 ka BP)*

		A-R	AO-R	AV-R	AOV-R	AOV-OV	AOV-AO	O	V	OV	f_o	f_v	f_{ov}	f_1	f_2	f_{12}
Boreal summer	NHL	1.7	1.2	2.2	2.5	1.7	1.3	−0.5	0.5	0.8	−0.29	0.29	0.47	−0.42	0.23	0.51
	NH	1	0.6	1.3	1.6	0.9	1	−0.4	0.3	0.7	−0.40	0.30	0.70	−0.67	0.23	0.81
	SH	0.2	−0.2	0.3	0.8	−0.1	1	−0.4	0.1	0.9	−2.00	0.50	4.50	2.00	0.33	−158
Boreal winter	NHL	−0.8	−0.5	−0.7	0.4	−0.4	0.9	0.3	0.1	0.8	−0.38	−0.13	−1.00	−0.60	−0.14	3.74
	NH	−0.5	−0.3	−0.4	0.6	−0.2	0.9	0.2	0.1	0.8	−0.40	−0.20	−1.60	−0.67	−0.25	2.75
	SH	−0.3	−0.2	−0.3	0.5	−0.2	0.7	0.1	0	0.7	−0.33	0.00	−2.33	−0.50	0.00	2.10
Annual average	NHL	0.2	0.1	0.5	1.2	0.4	1.1	−0.1	0.3	0.8	−0.50	1.50	4.00	−1.00	0.60	1.23
	NH	0.1	0	0.3	1	0.2	1	−0.1	0.2	0.8	−1.00	2.00	8.00		0.67	0.23
	SH	0	−0.1	0.1	0.7	0	0.8	−0.1	0.1	0.7				1.00	1.00	−1.00

Table 4.5 shows changes in near-surface air temperature in the NH, the northern hemisphere continents (NHL), and the SH between model simulations of the mid-Holocene climate and pre-industrial climate. $A - R$, $AO - R$, $AV - R$, $AOV - R$ represent the differences from the present-day climate R by using an atmosphere-only model, an atmosphere–vegetation model, an atmosphere–ocean model, and a fully coupled atmosphere–vegetation–ocean model, respectively. These data are taken from Ganopolski *et al.* (1998). According to the FS methodology O, V and OV depict the pure contributions owing to atmosphere–ocean interaction, to atmosphere–vegetation interaction, and to synergisms between these interactions respectively. $AOV - OV$ is the estimate of the full response of the system by linearly combining the pure contributions, but neglecting synergisms. f_o, f_v, f_{ov}, f_1, f_2, f_{12} are the amplification factors and feedback factors, respectively, as introduced in the text.

As discussed and stressed by Ganopolski *et al.* (1998), the atmosphere–ocean–vegetation (*AOV*) experiment offers a wealth of thought-provoking results. Unlike the other simulations, *AOV* reveals indeed a pronounced warming through the whole year in both hemispheres. In the NH, winter becomes warmer in spite of the lower insolation. This warming is caused by an extension of the boreal forests, a decrease of the subtropical deserts and a substantial reduction of sea ice, all contributing to a significant decrease of the planetary albedo. In the SH, the warming results from a weaker Atlantic thermohaline circulation, which feeds back negatively on the NH warming (the see-saw effect; Stocker, 1998). This weakening is due to an intensification of the hydrological cycle related to the warming and freshening of the North Atlantic and Arctic. Moreover, the comparison of *AVO* and *AV* shows that the boreal forests increased further and the subtropical deserts shrank further revealing strong synergetic effects between the different climatic subsystems. The aim of this chapter is to illustrate how to quantify the feedbacks and synergisms present in the paper by Ganopolski *et al.* (1998).

From Table 4.5, in which the Ganopolski *et al.* results with respect to temperatures are listed, it can be seen that the synergism *OV* due to interactions between the ocean and vegetation is always larger than the pure contributions *V* and *O*, the individual importance of the ocean and vegetation subsystems depending mainly upon the season under consideration.

Regarding the pure contribution of the ocean, it is expected that its thermal inertia will be responsible for damping any change of climate. Its largest impact is indeed in the NH, with a cooling of the boreal summer and a warming of the boreal winter leading to a small negative influence on the annual average in comparison with present-day climate. As can be seen from the negative sign of the amplification factor f_O, both the summer-time cooling as well as the winter-time warming oppose the pure reaction of the atmosphere as triggered by the insolation changes between mid-Holocene and present-day climate. However, the strong warming of the North Atlantic, together with an increased freshwater flux into the Atlantic basin, leads to a weakening of the thermohaline circulation. Due to the reduced oceanic heat transport, this in turn results in a warming of the SH (up to 2 °C near the Antarctic) and in a negative feedback for the NH high latitudes, which further damps the intensified seasonal contrast triggered by the insolation change.

The pure impact of vegetation, *V*, is mainly in the NH during the boreal spring and summer. As noted by Ganopolski *et al.*, the insolation-induced warming (finally up to 3 to 4 °C in the northern continental areas in winter and summer, respectively) is amplified due to a decrease of the surface albedo mainly in the snow-melting season. The latter is related to an extension of the boreal forests with their comparatively rough surface at the cost of the more smooth tundra as shown by Otterman *et al.*

(1984) and modeled by Gallée *et al.* (1992). As can be seen from the positive sign of the amplification factors f_V, the pure contribution of the vegetation amplifies the pure response of the atmosphere in the summer season and also in the annual mean – in contrast to the ocean.

The analysed pure impact of vegetation, with $V = 0.5\,°C$ over the NH continents, differs strongly from what one would conclude from a classical sensitivity experiment for the contribution of vegetation changes (given by $AOV - AO$ in Table 4.5). As demonstrated in Section 4.2.3, the difference comes from the contribution of the synergism $OV = 0.8\,°C$, which concerning temperature is the most important factor in all the cases shown, much larger than the pure contribution of the ocean or of the vegetation. It represents even the largest contribution to the response of the fully coupled model AOV, except for the northern latitudes during the boreal summer, where the atmosphere is, by itself, responsible for a warming of 1.7 and $1.0\,°C$ relative to present-day climate, respectively over land and over the whole hemisphere (against 0.8 and $0.7\,°C$ for OV). The latter fact can be seen more clearly from the large value of the amplification factor f_{OV}, the absolute value of which is greater than one except in summer in the NH. The positive sign of the factor also shows that the synergism is amplifying the pure atmospheric response in the summer season as well as in the annual mean, and its negative sign shows that it is opposing the atmospheric cooling during the boreal winter, just as for the pure contribution of the vegetation. The response of the fully coupled model without taking synergism into account ($AOV - OV$) clearly demonstrates the importance of the synergism here, as its value, $1.7\,°C$, substantially differs (by about $0.8\,°C$) from the response of the fully coupled model simulation $AOV = 2.5\,°C$.

As Ganopolski *et al.* (1998) state, in high northern latitudes a northward expansion of the boreal forests due to the summer warming in the fully coupled model leads to an annual warming via the vegetation–snow–albedo feedback, strongly amplified by the sea-ice–albedo feedback. The annual amount of sea ice indeed decreases as compared to the present day, which leads to the ocean absorbing more heat in summer and releasing it to the atmosphere in winter. Thus, it is the impact of the synergism that makes the vegetation contribution so important at the annual scale, as can be seen mainly from the negative values of $AOV - OV$ in winter.

In the subtropics, a strong positive feedback between vegetation and the stronger monsoon precipitation of the mid-Holocene leads to a greening of the Sahara with a vegetation cover of more than 80%. Contrary to the temperature, precipitation changes in summer over the NH continents (Table 4.6), North Africa in particular, are mainly related to the pure contribution of vegetation changes V and only to a small extend to the synergistic effects OV. As can be seen from the amplification

Table 4.6 *Precipitation in the atmosphere–ocean–vegetation models at the peak of the Holocene*

		A-R	AO-R	AV-R	AOV-R	AOV-OV	AOV-AO	O	V	OV	f_o	f_v	f_{ov}	f_1	f_2	f_{12}
Boreal summer	NHL	0.25	0.19	0.49	0.54	0.43	0.35	−0.06	0.24	0.11	−0.24	0.96	0.44	−0.32	0.49	0.36
	SHL	0.04	0.01	0.06	0.07	0.03	0.06	−0.03	0.02	0.04	−0.75	0.50	1.00	−3.00	0.33	3.10
	NAFRL	0.58	0.52	1.67	1.85	1.61	1.33	−0.06	1.09	0.24	−0.10	1.88	0.41	−0.12	0.65	0.15
Boreal winter	NHL	−0.01	−0.01	0	0.05	0	0.06	0	0.01	0.05	0.00	−1.00	−5.00	0.00		1.20
	SHL	−0.17	−0.13	−0.17	−0.01	−0.13	0.12	0.04	0	0.12	−0.24	0.00	−0.71	−0.31	0.00	−15.69
	NAFRL	−0.01	0	0.02	0.05	0.03	0.05	0.01	0.03	0.02	−1.00	−3.00	−2.00		1.50	−0.30
Annual average	NHL	0.08	0.06	0.19	0.26	0.17	0.2	−0.02	0.11	0.09	−0.25	1.38	1.13	−0.33	0.58	0.45
	SHL	−0.03	−0.03	−0.03	0.06	−0.03	0.09	0	0	0.09	0.00	0.00	−3.00	0.00	0.00	1.50
	NAFRL	0.18	0.16	0.63	0.72	0.61	0.56	−0.02	0.45	0.11	−0.11	2.50	0.61	−0.13	0.71	0.16

Explanation as in Table 4.5

factor f_V, especially in North Africa, the pure contribution of vegetation can be as much as two times larger than the pure atmospheric response. The amplification factors also demonstrate that, just as for the temperatures, the pure contributions of vegetation as well as of the synergism amplify the pure atmospheric response in summer and also in the annual mean, at least for the NH. At the same time, the ocean with its thermal inertia weakens this pure atmospheric response in both hemispheres.

In addition to the FS methodology, the feedbacks f_i and synergisms f_{ij} of the extended feedback analysis were also calculated. Their sign is actually the same as the sign of the additive amplification factors, except f_1 for temperature during the boreal summer in the SH and f_2 for precipitation during the boreal winter in North Africa. Their magnitude can be easily interpreted with, for example, vegetation explaining 23% of the change between the climate of the atmosphere–ocean system 6 ka ago and the present-day climate in the NH during boreal summer, f_{12} being three times as large. However, for the synergism we must take into account the additional correction terms $f_1(AOV - AO) + f_2(AOV - AV)$, which with $f_{12}(AOV - R)$ gives OV (see Eq. 4.15c). If the differences between AOV and AO or AV are small in comparison with $AOV - R$, f_{12} will provide the percentage of $(AOV - R)$ due to the synergism OV.

Thus, although derived from different assumptions, the FS methodology and the extended feedback analysis lead to comparable results. Results of the extended FS methodology complete those results because the additive amplification factors take the initial atmospheric response $(A - R)$ into account, helping to explain differences in the signs. For example, in the case of temperature changes in NH during boreal summer (Table 4.5), negative f_O and values smaller than one indicate that the interaction between ocean and atmosphere $(AO - A)$ opposes but does not overturn the response of the atmosphere-only model (from the value of $A - R$ it can be seen that, here, the warming relative to present-day climate, $AO - R$, is reduced). The same holds for the negative value of f_1, but in this case the conclusion is that the atmosphere–ocean interaction does not overturn the response of the atmosphere–ocean system. For the boreal winter, f_O and f_1 have the same negative sign (as also O) because the pure contribution of the ocean (a warming) is opposite to the reaction (a cooling) of both the atmosphere-only $(A - R)$ and the atmosphere–ocean system $(AO - R)$. For the SH during boreal summer, the positive value of f_1 as opposed to the negative value of f_O leads to the interpretation that, although the model that includes the ocean–atmosphere feedback yields lower temperatures than the atmosphere-only model $(AO - A$ is negative), the atmosphere-only remains warmer than the present-day reference, but the atmosphere–ocean system gets cooler.

4.4   Limitations of the techniques

The large values of f_{OV} for the case of annual averaged temperatures in the NH and mainly in the SH given in Table 4.5 are related to the fact that $A - R$ is small and even close to zero – which points to a potential problem of singularities when using the amplification factors of the extended FS methodology. The same holds for f_1 if $AO - R$ becomes zero. Moreover, f_{12} alone does not represent the pure contribution of the synergism, which is obtained by adding the contributions of f_1 and f_2 when applied to the difference between the fully coupled run and the respective simulation (see equation (4.15c)).

One point that is important for all the discussed methods remains the number of processes to be included in one particular analysis. In a non-linear system where most of the variables are interacting with each other, the real pure contribution of each component would require all the variables to be taken into account. For n variables, this would lead to 2^n experiments to be performed and a lot of synergisms involving 2, 3, …, $n-1$ and n variables. In our example, the problem is more simple because only two components (ocean, vegetation) were considered, leading to three contributions (pure ocean, pure vegetation and their synergism). However, the selection of factors is subjective. If a limited number of factors are selected, then the contributions of these factors include their pure contribution, and also their synergisms with the other factors that are not explicitly considered. For example, the atmosphere–vegetation feedback discussed in this paper implicitly includes feedbacks between the surface albedo, evaporation and atmospheric heat fluxes, and our analysis does not differentiate between their contributions. The presented methods are therefore not producing the full and final solution to the complex determination of the contribution of all processes involved and of all their interactions with the others, but they help considerably to start quantifying their importance objectively.

Finally, it is worth mentioning that the results of any feedback analysis depend on the ad hoc definition of feedbacks. Hence, a prerequisite of all feedback analyses is a careful definition of the meaning of the feedback under consideration. Furthermore, the actual value of a feedback or a factor depends on the definition of the calendar. If the same calendar for present and past climate is used, the factors of the annual mean contribution are the mean of the factors of the seasonal contribution. This is not the case if the astronomical definition of the calendar is chosen (Otto *et al.*, 2009a).

4.5   Conclusions

The presented techniques allow us to identify clearly the individual contribution of each process, including the feedbacks and synergisms. This helps to clarify the role they play in the climate system. In particular, let us comment on the seeming paradox that winters during mid-Holocene were milder than today in many regions of the NH, although the radiative forcing alone would cause a cooling. Both the FS methodology and the classical feedback analysis show positive feedbacks O and V. As the negative signs of the amplification factors f_O and f_V (Table 4.5) of the extended FS methodology and of f_1 and f_2 of the extended feedback analysis show, both the atmosphere–ocean feedback and the atmosphere–vegetation feedback tend to 'oppose' winter-time cooling occurring in both the atmosphere-only and in the atmosphere–ocean systems. However, they are not strong enough to produce a net warming and the response of the system without synergisms ($AOV - OV$) would produce a cooling with respect to today's climate. It is the synergism OV between these feedbacks that produces a winter-time warming – indicated by its large positive value. The large negative value of f_{OV}, much larger in absolute value than f_O and f_V, demonstrates that OV is opposed (a warming) to the strict reaction of the atmosphere (a cooling). For f_{12}, its large positive value is more difficult to explain. We need to go back to the sum

$$f_1 + f_2 + f_{12} = \frac{AOV - A}{AOV - R}$$

which expresses how much the reaction of the full system when compared to A is larger than its reaction when compared to the reference. Here f_1 and f_2, like f_O and f_V, are negative, the warming due strictly to the ocean and the vegetation being opposed to the cooling $A - R$, $AO - R$ and $AV - R$. f_{12} must therefore be much larger than the absolute value of $f_1 + f_2$ leading to a total contribution of O, V and mainly the synergism OV ($O + V + OV = AOV - A$) being two to three times larger than the climate change ($AOV - R$) between 6 ka BP and the present day. Hence the 'paradox' is only apparent, resulting rather from the synergism between vegetation and the ocean, involving sea ice in particular.

These results are valid for the CLIMBER-2.1 model used here. New simulations using a comprehensive climate system model indicate weaker contributions due to vegetation dynamics and weaker synergy (see Otto *et al.*, 2009b; Claussen, 2009). Work in progress suggests that the strength of factors and feedbacks might vary due to slow variability of ocean and sea-ice dynamics. Hence, a comprehensive FS or feedback analysis should include an assessment of the statistical robustness of the results obtained.

Finally, the decision on which method (including its advantages and disadvantages) to use then mainly depends on whether the user wants to make the point on (i) the contributions of the single feedbacks and synergisms to the final result or (ii) the amplification of the initial signal due to feedbacks and synergisms.

The question of which method one should apply depends on what is the aim of showing such analyses. The FS methodology shows the contributions (O, V, OV) by which the ocean, vegetation and their synergism impact on the final response of the fully coupled model in terms of units of the original signal A.

The additive amplification factors (f_O, f_V, f_{OV}) of the extended FS methodology demonstrate the importance of ocean and vegetation feedbacks and of synergism not only relative to each other, as done by O, V and OV, but also with respect to the response of the system without any feedback, i.e., the response of the pure atmosphere $(A - R)$: (i) The positive or negative signs of the amplification factors show whether the initial signal is amplified or weakened and (ii) absolute values larger or smaller than unity show whether the impact of ocean, vegetation or synergism is larger or smaller than the initial signal and thus what part is the most important in the final response of the fully coupled model. As discussed above, a problem with this method could arise from singularities in the cases of a vanishingly small normalisation factor $(A - R)$.

Feedbacks f_1, f_2 derived from the extended feedback analysis depict pure contributions of a given variable $(AO - A, AV - A)$ to its value in the respective simulation $(AO - R, AV - R)$ (and not in the fully coupled run or in the atmosphere-only). Only the contribution of the synergism has to be represented by a mixture of all factors and not only by f_{12} as discussed above. The sum of these feedbacks yields the difference between the full signal AOV and the signal A obtained without any feedbacks and synergisms relative to $AOV - R$ (see equation (4.17c)).

Acknowledgments

This work is supported by the European Research Council Advanced Grant EMIS (No. 227348 of the programme 'Ideas'). Q.Y. is supported by the Belgian National Fund for Scientific Research (F.R.S–FNRS).

5

Meso-meteorology: Factor Separation examples in atmospheric meso-scale motions

P. Alpert

Here, we present three examples of the Alpert–Stein Factor Separation Methodology (hereafter, FS in short) on a medium scale in the atmosphere, often referred to as meso-scale or, in general, meso-meteorology. The first example is that of a deep Genoa cyclogenesis (Alpert *et al.*, 1996a, b) that was observed during the Alpine Experiment (ALPEX) in March 1982, and then studied intensively by several research groups. The second example is that of a small-scale shallow short-lived meso-beta-scale – only tens of kilometers in diameter – cyclone over the Gulf of Antalya, Eastern Mediterranean (Alpert *et al.*, 1999). The third example is that of a much smaller scale, of orographic wind, following Alpert and Tsidulko (1994). In each of these three examples, some factors relevant to the specific problem are selected, and special focus is given to the role played by the synergies as revealed by the FS approach.

5.1 A multi-stage evolution of an ALPEX cyclone: meso-alpha scale

A relatively large number of studies have been devoted to cyclogenesis, with particular attention given to the processes responsible for the lee cyclone generation. Early studies of lee cyclogenesis (henceforth LC) focused on observations, and indicated the regions with the highest frequencies (Petterssen, 1956).

More recently, several theories have been advanced to explain the LC features, and they are frequently separated for convenience into two groups, as follows: the modified (by the lower boundary layer) baroclinic instability approach, as reviewed by Tibaldi *et al.* (1990) and Pierrehumbert (1985), and the directional wind shear suggested by Smith (1984).

Along with the theoretical approaches, numerical models have very frequently been used, starting with the pioneering work in the Alps by Egger (1972), in order

Factor Separation in the Atmosphere: Applications and Future Prospects, ed. Pinhas Alpert and Tatiana Sholokhman. Published by Cambridge University Press. © Cambridge University Press 2011.

to explain real cyclogenesis developments, as well as to verify the aforementioned simplified theories. These models have more recently become the most realistic tools, particularly with the full physics incorporation in the 3-D meso-scale models. Current numerical models seem to be very efficient in performing excellent simulations of complex LC developments. At the same time, the simplified theoretical models are unable to really clarify the complex physical mechanisms for the LC (Egger, 1988). Several studies of Alpine LC, both observational (Buzzi and Tibaldi, 1978), and modeling (Tosi *et al.*, 1983; Dell'Osso and Rabinovic, 1984) clearly indicate the existence of two distinct phases in LC development. The first is a rapid one, typically less than 12 hours, that was attributed to the blocking of the low-level cold front. The second deepening phase is a slower one (24 h), and better fits the typical growth rates of baroclinic instability (Tibaldi *et al.*, 1990).

Numerical simulations (Dell'Osso and Rabinovic, 1984) indicate that latent heat release is another major cyclogenetic contributor at the second phase. Other important mechanisms, particularly with regard to Alpine LC, were suggested to be fluxes of latent and sensible heat (Emeis and Hantel, 1984). However, no numerical simulations were attempted in order to separate all the aforementioned contributions at different stages. One of the major reasons, besides the modeling complexity, is the existence of many synergistic interactions among these processes that complicate the sensitivity study. Stein and Alpert (1993) (henceforth SA) have shown that for n factors being investigated, 2^n simulations are necessary in order to separate the contributions and all the possible interactions.

Alpert *et al.* (1996) applied the FS methodology to the most impressive Alpine LC development observed on 3–6 March 1982, during the Alpine Experiment (ALPEX). Quite a number of papers, mostly observational and through modeling, have dealt specifically with this case, and the comparison of their results with those obtained here by the FS method illustrates the power of this method. Furthermore, the choice of four factors: topography, convection, latent, and sensible heat fluxes, allow a direct comparison with the aforementioned studies. Another important mechanism is the potential vorticity (PV) advection that was found to be the dominant mechanism in some Alpine lee cyclogenetic cases (Bleck and Mattocks, 1984). It was not chosen as a factor here due to the complexity in switching the PV advection on/off, but was presented for the same event elsewhere (Tsidulko and Alpert, 2001), and indeed showed a most significant contribution by the PV/topography synergy.

5.1.1 The 3–6 March 1982 lee cyclone development

One of the most studied LCs, both observationally (Buzzi *et al.*, 1985, 1987) and by numerical simulations (Tibaldi and Buzzi, 1983; Bleck and Mattocks, 1984;

Dell'Osso, 1984; Tafferner and Egger, 1990), is the 3–6 March ALPEX case. A summary of some of the published literature and main objectives was given by Alpert *et al.* (1996b, Table 1). It was the most intense (Buzzi *et al.*, 1985) LC deepening during the ALPEX international special observational period, which was aimed at a better understanding of the role of mountains in weather. Figures 5.1A and B show the 4 and 5 March 1200 UTC analyzed surface maps based on ECMWF (European Centre for Medium Range Weather Forecasts) initialized analyses. An 8.7 hPa (mb) deepening in the lee of the Alps occurred within 24 hours, with most of the pressure fall (6.1 hPa) occurring within the last 12 h period. Our 48 h model simulation is shown in Fig. 5.1C, and its verifying analysis in Fig. 5.1B.

Following earlier studies, four factors were chosen: topography (t), surface latent heat flux (l), surface sensible heat flux (s), and the latent heat release (r) in the deep clouds developing within the cyclonic system. In order to calculate all the possible 16 (2^4) contributions, meso-scale simulations with the PSU/NCAR meso-scale model version MM4 were performed. Horizontal resolution was 80 km with a 46×34 mesh and 16 levels up to 16 km. It should be pointed out that in the early 1990s, 80 km was still referred to as meso-scale. Further details on model and simulations can be found in Alpert *et al.* (1996), and in SA. The initial time for the simulations was 4 March 1200 UTC. The sea-level pressure change at the center of the control-run cyclone was then partitioned into contributions by each process or combination of processes. These included four "pure" contributions (t, l, s, and r), six double interactions (tr, ts, sr, lr, tl and sl), four triple interactions (tsl, tsr, tlr, and slr), and even one quadruple interaction ($tslr$). The residual was referred to as the "large-scale contribution". This term is excluded from the following discussion because only the local processes were investigated.

Figure 5.2 presents the time evolution for all the 15 local contributions, illustrating the dominant processes at each stage for the Genoa cyclone deepening. Figure 5.2A focuses on the five major contributions to be discussed later in more detail, while Figs. 5.2B and C show the rest of the contributions. The mechanical effect of the Alpine topography (t) is the first to generate the rapid deepening within 6 h (Fig. 5.2A), while all other contributions are still much smaller. This corresponds well to the aforementioned first deepening phase. In the second phase, the latent heat release (r) or "convection only" becomes the dominant contributor (42–54 h), while the "pure" topographic contribution quickly diminishes to become later a major cyclolytic (destruction of cyclone) factor. This is probably due to the cyclone's motion beyond the favorable lee region. Of interest here is the considerable contribution of the synergistic double and triple interactions. For example, the mountain-induced convection (tr) is the second contributor at the first stage (27–36 h). Each of the synergistic terms can be associated with a specific meaning, thus shedding light on the complex physical mechanisms under investigation.

P. Alpert

Figure 5.1 Surface pressure charts for the ALPEX lee cyclogenesis on 3–6 March 1982. (A) 4 March 1200 UTC ECMWF analysis; (B) 5 March 1200 ECMWF analysis; (C) 48 h model prediction for the model prediction in (B). The isobaric interval is 2.5 hPa. The mesh is due to plotting software.

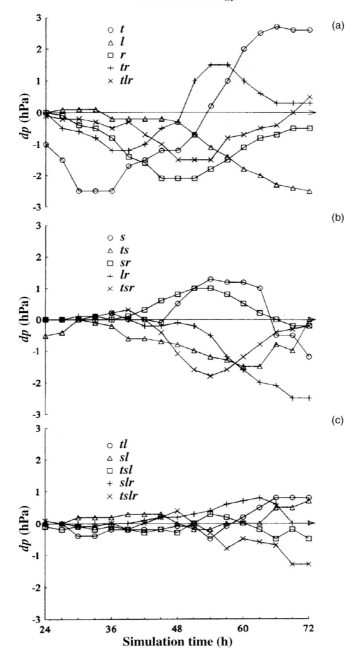

Figure 5.2 The 48 h time evolution for all the 15 local contributions to the ALPEX cyclone deepening (hPa) following 24 h of simulation. The interaction synergistic contributions are indicated by the letters of their parent factors. (A) First five major contributors; (B) next five contributors; (C) last five contributors.

Table 5.1 *Primary and secondary cyclogenetic processes as well as major cyclolytic contributors and their time evolution, based on the factor separation for sea pressure at the cyclone's center*

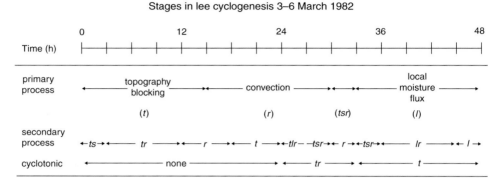

Stages in lee cyclogenesis 3–6 March 1982

tlr : mountains convection induced by local moisture fluxes.
tsr : mountains convection induced by sensible heat fluxes.
ts : mountains heating.
tr : mountains convection.
lr : convection due to local moisure fluxes.

The triple interaction **tlr**, found to be quite important at the 45–60 h phase, for example, represents the contribution to the deepening by the terrain-induced convection with local moisture, which is the triple synergism of terrain, convection, and surface latent heat flux. Table 5.1 summarizes the primary and secondary cyclogenetic processes, as well as the major cyclolytic contributors that emerge from the surface pressure analysis. Obviously, the results here, as well as in the other examples, are valid only to the extent that the models correctly simulate the real atmosphere.

5.1.2 Absolute comparisons for effects of several processes

Earlier, four factors were chosen. Alpert *et al.* (1995) link the sensitivity of the FS results to the number and type of factors to be selected. They show that the increase in the number of factors diminishes the individual contribution of a particular factor. If, for example, topography, which is a crucial factor, is included in each of the eight potential sets $\{t\}$, $\{t, s\}$, $\{t, l\}$, $\{t, r\}$, $\{t, s, l\}$, $\{t, s, r\}$, $\{t, l, r\}$ and $\{t, s, l, r\}$, its contribution varies significantly from case to case as shown in Fig. 5.4. Notice that the notation with { } refers to a potential numerical experiment for which the chosen factors or processes to be tested are listed within the parentheses as a group. This notation should not be confused with that of a specific contribution isolated for a particular experiment. For example, the triple interaction **tlr** in the

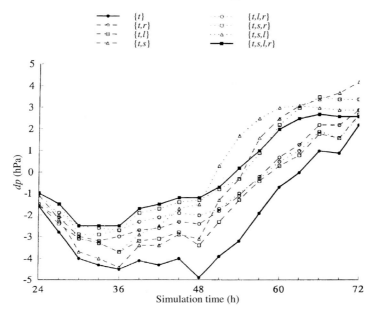

Figure 5.3 The pure topographic contribution (*t*) (hPa) to the pressure fall at the center of the control cyclone as a function of the simulation time. Each curve represents one of the eight possible sets of factors (see text). Heavy lines indicate the maximum choice of the one factor only{*t*} and the maximum choice of the four factors, i.e., {*t, s, l, r*}.

preceding section was isolated in an experiment for which four factors were chosen, i.e., the set {*t, s, l, r*}. Of course, each set is a legitimate choice for a modeler who investigates the role of topography (*t*) in the lee cyclogenesis, and such examples are found in the literature (see SA). In Section 5.2, we have chosen the specific set with all four factors included, i.e., {*t, s, l, r*}. Figure 5.3 presents the "pure" topographic contribution to the cyclone deepening for each of the eight possible sets of factors to be investigated. It is not unexpected that the topographic contribution is largest when only one factor {*t*} was considered, and the smallest, at least until the 48th hour of simulation was reached, when the largest set of factors {*t, s, l, r*} was chosen, since, as the number of factors increases, synergistic contributions between the new factors and topography are extracted out from the original *t* contribution. In the extreme case where only topography was chosen as a factor, all synergistic contributions with the topography are tacitly assumed to be part of the topographic contribution, making it the largest – see the bottom curve in Fig. 5.3. For example, the cyclogenetic contribution of the latent heat release synergism with topography (*tr* in Fig. 5.2A) is associated with the topography for the case when the factor *r* is not an investigated factor.

5.1.3 Spread of the factor separation model simulations:
another by-product study tool

Earlier, an experiment with a set of four factors was described. The FS method therefore required $16 = 2^4$ different simulations with all possible combinations of factors switched on/off. Obviously, in each simulation, the cyclone center may have a different location, which reflects the specific dynamical evolution due to the contribution of factors switched on. Figure 5.4 presents the 16 different locations of the cyclone center, following 45 hours of simulation time. In this notation, the factors switched on are listed, so **TLR**, for instance, indicates the simulation where only the factors **t**, **l**, and **r** were operating, while **s** was switched off.

A remarkable feature in the spread presented in Fig. 5.4 is the tendency of the factors to attract the cyclone center toward different regions. The convection (**r**) drifts the cyclones to the east-northeast, while topography moves similarly but to the north-northwest, and the latent/sensible heat fluxes (**l** and **s**) tend to move the cyclone to the south, towards the sea. This feature is even more pronounced as the spread of the solutions with time is analyzed. Alpert *et al.* (1996b, Fig. 8) present

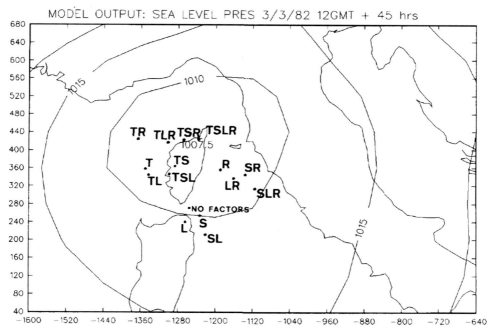

Figure 5.4 The 16 locations of the cyclone center following 45 h of simulation for the 16 simulations with all possible combinations of factors switched on/off. In this notation, the factors switched on are listed: hence, **TLR**, for instance, indicates the simulation where only the factors **t**, **l**, and **r** were operating while **s** was switched off. From Alpert *et al.* (1996b, Fig. 7)

the cyclone centers, as in Fig. 5.4, but at 30, 45, and 60 hours of simulation time. As time evolves, and the cyclone spread region moves away from the lee cyclogenetic area, at 60 hours, a clear change of the aforementioned tendencies takes place. The topographic factor delays the cyclone at the west over Sardinia, while sea fluxes move the cyclone to the east, where a larger and warmer body of water dominates. Alpert *et al.* (1996b) point out that, as the cyclone center's spread region increases further with time, the FS for a very specific point, such as the center of the control run, becomes less and less meaningful.

5.2 A shallow, short-lived meso-beta cyclone over the Gulf of Antalya, Eastern Mediterranean

The physical mechanisms of a shallow, short-lived meso-beta scale cyclone, i.e., order of a few tens of kilometers, over the Gulf of Antalya, Eastern Mediterranean, are studied, with the PSU/NCAR MM5 meso-scale model (Alpert *et al.*, 1999). The main results of this study follow. Although the thin stratus clouds within this cyclone as observed from satellites are not resolved even by the 3 km model nesting, the dynamical evolution and the 3-D structure, including also the increase of humidity at the center of the cyclone, were well captured. The small cyclone or eddy develops before sunrise following convergence of the strong katabatic winds from the nearby steep Anatolia mountain slopes that have 2 km high peaks. The eddy's lifetime is of the order of 5–7 h, and it quickly dissipates before noon. Based on the simulated vertical winds, vorticity, and humidity, as well as the observed IR cloud- top temperatures, the depth of the eddy is estimated to be 500–800 m. It is shown that the divergence term in the vorticity equation is dominant during the eddy's generation. A Lagrangian analysis for the trajectories of several air masses that were identified as crucial for the eddy's development reveals a sharp increase both in the PV (potential vorticity, by 7–8 units), and in the specific humidity (3.5 to 7 g kg^{-1}) as the air parcels descend from about 840 to 980 hPa. The air-parcel analysis also shows that the diabatic contribution is quite important.

The FS experiments confirmed that pure topography is the major factor, while the synergistic effect of sea fluxes and topography contributes about 20% of the total vorticity. The Antalya cyclone is common during July to September morning hours, and its frequency of occurrence was estimated from satellite pictures to be about 20%.

5.2.1 *The Antalya cyclone*

A zoom over the Gulf of Antalya, Fig. 5.5, reveals a beautiful meso-beta scale cyclone evolution. Figures 5.5a–c show three consecutive NOAA satellite pictures

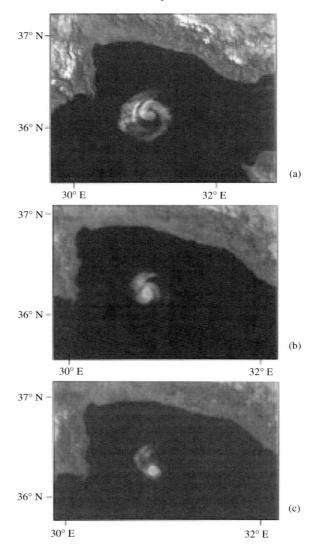

Figure 5.5 Satellite images over the eastern Mediterranean on 8 July 1993 with a zoom over the Gulf of Antalya. (a) NOAA10 NIR 0652 LT (0352 UTC); (b) NOA 12 visible 0826 LT (0526 UTC); (c) NOAA9 visible 0924 LT (0624 UTC). Full E–W horizontal scale of the picture is 250 km (a) and 200 km (b, c).

at 0652, 0826 and 0924 LT, respectively. The pictures are in the near IR, visible, and visible of NOAA10, NOAA12, and NOAA9, respectively. The cyclone was generated before sunrise, at around 0400 LT (sunrise at 0540 LT), and quickly dissipated after sunrise with a lifetime of about 5–7 h. Figure 5.5c is just before the eddy has completely dissipated.

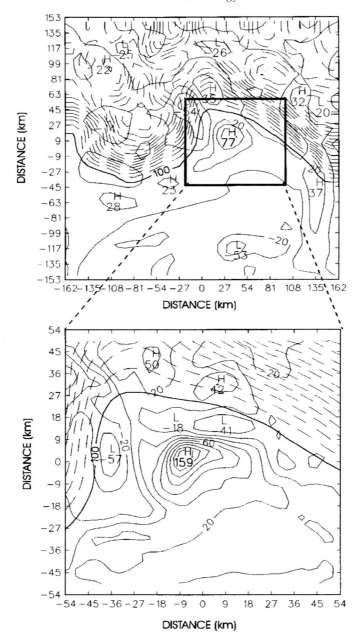

Figure 5.6 MM5 simulation. Model relative vorticity (10^{-5} s^{-1} units) in the coarse (upper panel) and in the nested (lower panel) domains at 0600 UTC. Contour interval for vorticity is 20 units. Topography is indicated by dashed lines with an interval of 100 m.

The peak of the simulated Antalya vortex is reached at 0600 UTC, and afterwards the eddy quickly dissipates, as also observed from the satellite pictures in the example in Figs. 5.5a–c. The maximum vorticity at the center of the eddy simulated in the nested model run reaches 159×10^{-5} s^{-1} (Fig. 5.6), as compared to 77×10^{-5} s^{-1} in the coarse grid and 83×10^{-5} s^{-1} only in the hydrostatic run. The maximum wind speed within a radius of about 30 km from the cyclone center is of about 6 m s^{-1}, increasing in the nested run to about 7 m s^{-1}.

5.2.2 Factor separation results

The contribution of the sea fluxes to cyclone development was calculated using the FS method. Here, the FS experiments confirmed that pure topography is the major factor, contributing 80% of the central cyclone's vorticity, while the synergistic effect of sea fluxes and topography contributes the remaining 20%. This value may in reality be somewhat greater, since the thin stratus clouds were not captured by the model.

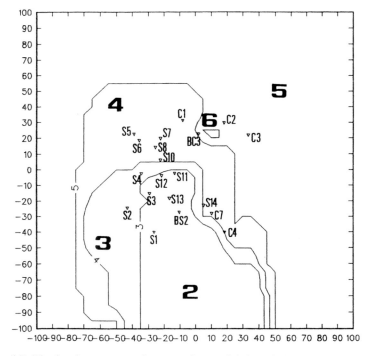

Figure 5.7 The land-use categories over the model domain based on NCAR classification over a grid of $(1/6° \times 1/4°)$. Station locations are indicated by triangles. Physical parameters for each category are listed in Table 5.1.

The 20% result is clearly different from the case with larger synoptic scale Cyprus cyclones over the Eastern Mediterranean, where the sea fluxes are the major contributor, as shown by Alpert *et al.* (1995).

5.3 Separation of thermal and terrain effects on mountain winds

Figure 5.7 shows the land-use categories over the model domain in the Sacramento valley, California, following meso-scale simulations performed by Alpert and Tsidulko (1994). These simulations were part of Project WIND, US Army, in which a large field operation and a meso-scale model inter-comparison study were undertaken (Pielke and Pearce, 1994). Details on the PSU-NCAR model simulation

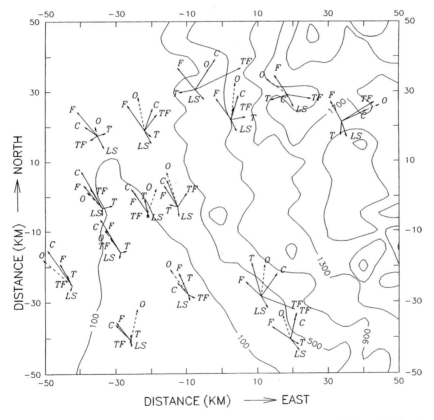

Figure 5.8 Model domain 100 km × 100 km in Sacramento Valley, California. Topographic contour interval is 400 m. Observed (**O**; dashed) wind vectors along with the control run result (**C**) for 1500 PST 27 June 1985 are plotted. Factor-separated wind vectors are topography (**T**), fluxes (**F**), topography–flux synergism (**TF**), and the residual large-scale (**LS**). The four contributions sum up to the control wind vector.

and land-use types and resolutions (horizontal 5 km to be discussed here) are given in Alpert and Tsidulko (1994). The surface stations for verification are indicated by the triangles in Fig. 5.7. Of special interest here are the results of applying the FS method to the surface wind vectors. Each vector in the control run was decomposed into the four following vectors; terrain (T), surface sensible heat fluxes (F), terrain-fluxes interaction (TF) and the residual vector attributed to the large-scale effect (designated as LS).

Figure 5.8 presents the factor-separated four vectors in all fifteen locations where the observed wind was available. The observed wind was added for comparison with the control (C) result and is denoted by the O vector. One interesting result is that the synergic TF contribution is the dominant one over the mountainous region – the NE section of the domain in Fig. 5.8 – whereas at the Sacramento Valley – the SW section – the surface flux, which is the F contribution, dominates. This is not surprising, since the TF synergic contribution represents the anabatic upward mountain wind that is generated by the combined action of the thermal fluxes and sloping terrain. In contrast, the pure terrain effect, T, represents only the mechanical mountain effect and is smaller at this time of day, i.e., 1500 local time (PST). This is an example of how the FS methodology provides a better physical insight into the different mechanisms responsible for the orographic wind. In this case it also provides a first approach to the quantification of the anabatic wind.

The three examples described in this chapter are all of different modeling simulations on the meso-scale, and they all demonstrate the better understanding of complex physical processes that is gained by the FS methodology. The variables analyzed in these three examples change from the surface pressure at the cyclone center in the first case to vorticity in the second case, and to surface wind vectors over steep topography. As expected, synergy is shown to be a major player in meso-scale processes, irrespective of the variable or factors chosen.

6

Using the Alpert–Stein Factor Separation Methodology for land-use land-cover change impacts on weather and climate process with the Regional Atmospheric Modeling System

A. Beltrán-Przekurat, R. A. Pielke Sr., J. L. Eastman,
G. T. Narisma, A. J. Pitman, M. Lei, and D. Niyogi

The Alpert–Stein Factor Separation Methodology (FS) has been applied effectively in the Regional Atmospheric Modeling System (RAMS) to assess the relative contribution of different factors to weather and climate processes. In this chapter we will discuss model sensitivities to historical and future changes in land-use land-cover (LULC), biophysical and radiative effects of increased carbon dioxide (CO_2) concentration and land-cover representation assessed using FS in weather and regional climate simulations for various regions around the world. This method emphasizes the importance of land-cover changes and CO_2 biological effects when addressing regional-scale future climate change impacts.

6.1 Introduction

Observations and modeling studies show that land-surface properties can influence the near-surface atmosphere through exchanges of heat, moisture, momentum, gases, and aerosols on timescales ranging from seconds to years, and on local to regional and possibly global spatial scales (Pielke, 2001; Arora, 2002; Pielke *et al.*, 2002; Pitman, 2003; Niyogi *et al.* 2004; Foley *et al.*, 2005). Urbanization, deforestation–reforestation, conversion of natural areas to agriculture, and increases in irrigation areas are some LULC modifications that often affect albedo, leaf area, roughness length, and root biomass. These landscape modifications can lead to changes in near-surface fluxes that affect temperature (e.g., Baidya Roy *et al.*, 2003; Strack *et al.*, 2008), humidity (e.g., Douglas *et al.*, 2006, 2009; Roy *et al.*, 2007), boundary-layer process, and precipitation (Pielke *et al.*, 2007a). These changes can potentially feed back to the biophysical variables through a two-way interaction,

Factor Separation in the Atmosphere: Applications and Future Prospects, ed. Pinhas Alpert and Tatiana Sholokhman. Published by Cambridge University Press. © Cambridge University Press 2011.

enhancing or decreasing the initial perturbation (Pitman, 2003; Pielke and Niyogi, 2008).

Increase in atmospheric carbon dioxide (CO_2) concentrations is another process that alters near-surface temperatures (Solomon *et al.*, 2007). The biological response to increases in CO_2 includes a decrease in stomatal conductance (g_s), an increase in carbon assimilation, biomass, and leaf area index (LAI) (Cao *et al.*, 2009). Changes in g_s and LAI affect the energy partition between sensible and latent heat, through the transpiration process, which in turn affects soil thermodynamics and eventually near-surface temperature and humidity (Niyogi *et al.*, 2009).

Complex nonlinear interactions between atmosphere and biosphere under changes LULC and increasing CO_2 levels occur at a wide range of spatial scales, global (e.g., Chase *et al.*, 2001) to regional (e.g., Eastman *et al.*, 2001), and temporal scales, decadal to monthly (e.g., Narisma and Pitman, 2004; Alpert *et al.*, 2006). A key aspect in the understanding of how those processes may affect the climate system is to be able to recognize the underlying mechanisms of the effects induced by each of the factors separately or by their interactions on near-surface atmospheric circulation. The Alpert–Stein Factor Separation Methodology (Stein and Alpert, 1993) can be used to assess the contributions of each of the factors separately and also their synergistic interactions.

In this chapter we review modeling studies that applied the FS methodology to assess the relative contribution and the interactions of:

- historical and future changes in land cover with biophysical and radiative effects of increased CO_2 concentrations (Eastman *et al.*, 2001; Beltrán, 2005; Pitman and Narisma, 2005; Narisma and Pitman, 2006);
- land-cover change patterns (Gero and Pitman, 2006); and
- improved boundary conditions data set and urban representation (Lei *et al.*, 2008).

The modeling experiments in those studies employed the Regional Atmospheric and Modeling System (RAMS; Pielke *et al.*, 1992; Cotton *et al.*, 2003), coupled with a plant model (GEMRAMS) and with an urban model (RAMS-TEB) in some of the experiments. These experiments use a wide range of modeling domains, atmospheric boundary conditions, and land-cover data sets.

This chapter is organized as follows. A brief description of RAMS, GEMRAMS, and RAMS-TEB is provided in Section 6.2. Applications of the Alpert–Stein FS methodology with RAMS and GEMRAMS for LULC changes and CO_2 concentration levels are presented in Section 6.3. Section 6.4 describes the application of the methodology to evaluate the effects of the spatial pattern of land-cover changes. Section 6.5 describes how the Alpert–Stein FS methodology can help to evaluate the relative contributions of model and data improvements. Final comments are presented in Section 6.6.

6.2 RAMS, GEMRAMS, and RAMS-TEB description

RAMS is a general-purpose, atmospheric-simulation model that includes the equations of motion, heat, moisture, and continuity in a terrain-following coordinate system. It is a fully three-dimensional and non-hydrostatic model. RAMS has been applied to a diverse range of spatial and temporal scales, to study a wide range of processes over domains located all over the world (e.g., Lawton *et al.*, 2001; Cotton *et al.*, 2003 and references therein; De Wekker *et al.*, 2005; Narisma and Pitman, 2004; Beltrán, 2005 and references therein; Douglas *et al.*, 2006, 2009; Roy *et al.*, 2007). RAMS contains several options for the typical physical parameterizations of atmospheric modeling systems. Table 6.1 shows the options available in the version of the RAMS used in the examples presented in this chapter.

To analyze the effects of CO_2 on near-surface atmosphere most of the studies discussed in this chapter used the fully coupled atmosphere–plant model version of

Table 6.1 *Summary of the parameterization options available in RAMS, GEMRAMS, and RAMS-TEB*

Parameterization	Options	References
Turbulence	Mellor–Yamada; anisotropic deformation; isotropic deformation; Deardorff	Smagorinsky (1963), Mellor and Yamada (1982), Cotton *et al.* (2003)
Radiation	Mahrer–Pielke; Chen–Cotton; Harrington	Mahrer and Pielke (1977), Chen and Cotton (1983), Harrington (1997)
Clouds microphysics	No cloud; condensation only; simple dump-bucket; single-momemtum	Rhea (1978), Cotton *et al.* (1995), Walko *et al.* (1995)
Convective parameterization	Modified Kuo; Kain–Fritsch	Molinari (1985), Tremback (1990), Kain (2004), Castro *et al.* (2005)
Surface layer	Specified surface layer gradients; LEAF-2	Walko *et al.* (2000)
Vegetation dynamics	Seasonal and latitudinal leaf area index (LAI) dependence (from climatology); LAI from satellite (climatology or current); prognostic LAI based on daily net carbon assimilation (i.e., GEMRAMS)	Walko *et al.* (2000); Lu and Shuttleworth (2002); Eastman *et al.* (2001); Beltrán (2005); Beltrán-Przekurat *et al.* (2008)
Urban model	RAMS-default urban land cover; TEB	Walko *et al.* (2000); Rozoff *et al.* (2003)

RAMS that is called GEMRAMS, comprising the RAMS and the General Energy and Mass Transport Model (GEMTM; Chen and Coughenour, 1994, 2004), an eco-physiological process-based model. One of the main features of GEMRAMS is that the atmosphere and biosphere are allowed to dynamically interact through the surface and canopy energy balance. Vegetation grows as a function of temperature, radiation, and the water status of the soil and atmosphere. Leaf area index (LAI), computed from leaf biomass, and canopy conductance, estimated from interactions between transpiration, photosynthesis and root-water uptake, are the variables connecting the carbon and energy flux components.

GEMTM and the soil–vegetation–atmosphere transfer scheme in RAMS, the Land Ecosystem–Atmosphere Feedback model version 2, LEAF-2, (Walko *et al.*, 2000), connected through LAI and canopy conductance, represent the storage and exchange of heat, moisture, and carbon associated with the near-surface atmosphere and biosphere. At each time step, photosynthesis at the leaf level is calculated for sunlit and shaded leaves and also separately for C_3 and C_4 species (Farquhar *et al.*, 1980; Chen *et al.*, 1994, 1996) as a function of photosynthetic active radiation, temperature, and CO_2 concentration. Stomatal conductance for sunlit and shaded leaves is computed using the semi-empirical linear Ball–Berry relationship based on net photosynthesis, relative humidity, and leaf surface CO_2 (Ball *et al.*, 1987). Leaf photosynthesis and conductance are scaled up to the canopy level using sunlit and shaded LAI (Chen and Coughenour, 1994; de Pury and Farquhar, 1997), calculated using light extinction coefficients from a multi-level canopy radiation model (Goudriaan, 1977). The available photosynthate is allocated to leaves, stems, roots, and reproductive organs with variable partition coefficients, which are functions of soil water and temperature conditions. Daily, a new total LAI value is estimated from the daily leaf biomass growth using the vegetation-prescribed specific leaf area.

To study the effects of urbanization on heavy rainfall and surface temperature, Lei *et al.* (2008) used a version of RAMS coupled to an explicit urban model, the Town Energy Budget (TEB) (Masson, 2000; Rozoff *et al.*, 2003). TEB follows a generalized urban canyon approach, considering energy budgets separately for roofs, walls, and roads, and also the radiation interactions between those surfaces (Rozoff *et al.*, 2003). Anthropogenic heat release is also incorporated into the TEB.

6.3 LULC and carbon dioxide changes

Applications of the Alpert–Stein FS methodology with RAMS and GEMRAMS to study the contributions of LULC and CO_2 changes are given in Eastman *et al.* (2001), Beltrán (2005), Pitman and Narisma (2005), and Narisma and Pitman (2006).

Table 6.2 *Summary of the experiments performed (Exp), the difference fields and their meaning in Eastman et al. (2001) and Beltrán (2005)*

Exp	Land cover CURR	NAT	CO_2 radiation CO_2	$2\times CO_2$	CO_2 biology CO_2	$2\times CO_2$	Differences	Contributions from:
f0	✓		✓		✓			
f1		✓	✓		✓		$\hat{f}1 = f1 - f0$	Natural vegetation
f2	✓			✓	✓		$\hat{f}2 = f2 - f0$	$2 \times CO_2$ radiation
f3	✓		✓			✓	$\hat{f}3 = f3 - f0$	$2 \times CO_2$ biology
f12		✓		✓	✓		$\hat{f}12 = f12$ $- (f1 + f2) + f0$	Natural vegetation and $2 \times CO_2$ radiation
f13		✓	✓			✓	$\hat{f}13 = f13$ $- (f1 + f3) + f0$	Natural vegetation and $2\times CO_2$ biology
f23	✓			✓		✓	$\hat{f}23 = f23$ $- (f2 + f3) + f0$	$2 \times CO_2$ radiation and $2 \times CO_2$ biology
f123		✓		✓		✓	$\hat{f}123 = f123$ $- (f1 + f2 + f3)$ $+ (f12 + f13$ $+ f23) - f0$	Natural vegetation and $2 \times CO_2$ radiation and $2 \times CO_2$ biology

6.3.1 Sensitivity to past LULC and carbon dioxide concentration using current atmospheric conditions

Eastman *et al.* (2001) considered the effects of three factors, LULC and radiative and biophysical CO_2 levels, on near-surface atmospheric variables during the 1989 growing season, April to October. The sensitivity experiments were performed over a simulation domain centered on the central Great Plains of the United States. Table 6.2 summarizes the modeling experiments. The land-cover experiment $\hat{f}1$ corresponds mainly to a conversion from tall and short grasslands to croplands, representing the changes in land use since pre-settlement to the present. In the CO_2 biology experiment only the CO_2 levels that affect physiological processes, such as photosynthesis and respiration parameterized within GEMTM, were allowed to change. The CO_2 effect on the radiative fluxes corresponded to the CO_2 radiation experiment (Table 6.2). To isolate the effects of LULC and CO_2 levels, all the simulations used the National Centers for Environmental Prediction–National Center for Atmospheric Research (NCEP-NCAR) reanalysis (Kalnay *et al.*, 1996) as atmospheric boundary conditions.

A. Beltrán-Przekurat, et al.

Table 6.3 *Summary of the seasonal and domain-averaged daily maximum (Tmax) and minimum temperature (Tmin) and precipitation and contributions from all the factors in the Eastman et al. (2001) sensitivity experiments (see also Table 6.2). Relative changes (×100) with respect to the control experiment (f0) are shown in parentheses.*

		Tmax (°C)	Tmin (°C)	Precip (mm d^{-1})
Control experiment	f0	23.13	7.30	0.89
Differences	$\hat{f}1$	−1.19 (5)	−0.02 (<1)	−0.04 (5)
	$\hat{f}2$	0.01 (<1)	0.10 (1)	0.01 (1)
	$\hat{f}3$	−0.75 (3)	0.26 (4)	−0.05 (6)
	$\hat{f}12$	0.03 (<1)	−0.01 (<1)	~0
	$\hat{f}13$	0.07 (<1)	0.07 (1)	~0
	$\hat{f}23$	0.02 (<1)	~0	~0
	$\hat{f}123$	0.01 (<1)	~0	~0

The results showed that the land-cover and biological CO_2 changes were the dominant factors in the sensitivity experiments (Table 6.3) over the seven months simulated. For all the variables analyzed, the contribution of the radiative CO_2 changes was minimal. Table 6.3 shows the seasonal domain-averaged contribution of each of the factors and their interaction to daily maximum (Tmax) and minimum (Tmin) temperature and precipitation (Precip). In the case of Tmax, the contributions of the natural vegetation ($\hat{f}1$) and $2 \times CO_2$ biology ($\hat{f}3$) by themselves indicate a large cooling effect of 1.19 °C and 0.75 °C , representing a 5% and a 3% change with respect to the control experiment (i.e., current land cover and CO_2 levels). The $2 \times CO_2$ biology ($\hat{f}3$) had the largest effect on Tmin, with an average positive change of 4% with respect to the control experiment.

The interactions among the factors were relatively small in most of the variables, but in the case of Tmin, one of them, the contribution from natural vegetation and $2 \times CO_2$ biology ($\hat{f}13$), was of the same magnitude as the pure $2 \times CO_2$ radiation effect. Although the $\hat{f}13$ value was small (1% of positive change), this is an indication that the biological effects of increasing CO_2 levels synergistically interact with land-cover changes to affect Tmin: the warming due to the reduction of the long-wave loss, caused by an elevated near-surface atmospheric water vapor, is compensated by the slight cooling associated with the natural vegetation.

Beltrán (2005) addressed the total LULC and CO_2 changes for a simulation domain covering southern South America (only experiments f0 and $\hat{f}23$ were carried out). Here, we present results of additional simulations that complete the set of experiments required by the Alpert–Stein FS methodology, similar to the modeling approach of Eastman *et al.* (2001) (Table 6.2). The simulation experiments were performed for the 1996–7 austral spring season, September to November.

The lower atmospheric boundary conditions were provided by the European Centre for Medium-Range Weather Forecasts Reanalysis ERA-40 (Uppala *et al.*, 2005). Sensitivity experiments were performed using the current vegetation cover (Fig. 6.1a) and a "historical" one that represents the conditions before European settlement. Croplands were replaced by tall grass, wooded grasslands, or ever-green broadleaf forest, depending on their geographical location based on the vegetation maps of Matthews (1983), Küchler (2000), and Cabrera and Willink (1980) (Fig. 6.1b). As in Eastman *et al.* (2001) two CO_2 concentration val-ues, 360 ppm and 720 ppm, were used to address the biological and radiation effects and interactions of increasing CO_2 (Table 6.2) on near-surface weather variables.

In a domain-averaged basis, the pure and interaction effects in these sensitivity simulations were almost negligible. Nevertheless, spatial differences were notice-able. The contributions of LULC changes ($\hat{f}1$) and increased CO_2 values in the biophysical processes ($\hat{f}3$) showed the largest effect on the analyzed variables, mostly associated with changes in near-surface latent and sensible heat flux. For example, for the October–November period, Tmax was up to 1.2°C higher with a natural vegetation landscape than with the current vegetation in areas where vegeta-tion changed from C_3 grasslands to wheat and soybean (i.e., in central and southern Pampas), and was up to 0.3°C lower when crops replaced evergreen trees (e.g., in southern Brazil) (Fig. 6.2). The contribution from the CO_2 biophysical changes ($\hat{f}3$) showed Tmax increases of up to 0.4 °C in areas associated with an increase in LAI values (Fig. 6.2). Interestingly, the interaction term $\hat{f}13$ that represents the contributions from natural vegetation and $2\times CO_2$ biology has a negative effect on Tmax in the southern Pampas. Tmax is still higher in the $\hat{f}13$ experiment (i.e., nat-ural vegetation and $2\times CO_2$ biology) with current vegetation, f0, but the combined effect on Tmax is lower than the pure natural vegetation ($\hat{f}1$) and $2 \times CO_2$ ($\hat{f}3$) separate contributions (Fig. 6.2). The other interaction terms ($\hat{f}12$, $\hat{f}23$ and $\hat{f}123$) have negligible contributions.

The effect of increased CO_2 concentration was less noticeable on precipitation, except in localized areas in the northern part of the simulation domain. In Brazil, precipitation was up to 1mm day^{-1} larger under the natural landscape (i.e., ever-green trees) than under the current agricultural (i.e., corn) conditions (Fig. 6.3). This area of increased rainfall can still be seen in the contributions due to the interaction term $\hat{f}123$ (Fig. 6.3). This is an indication of the complex mechanisms and fac-tors that can be involved in detecting effects on precipitation under future climate conditions (i.e., land-use and higher CO_2 concentrations).

A. Beltrán-Przekurat, et al.

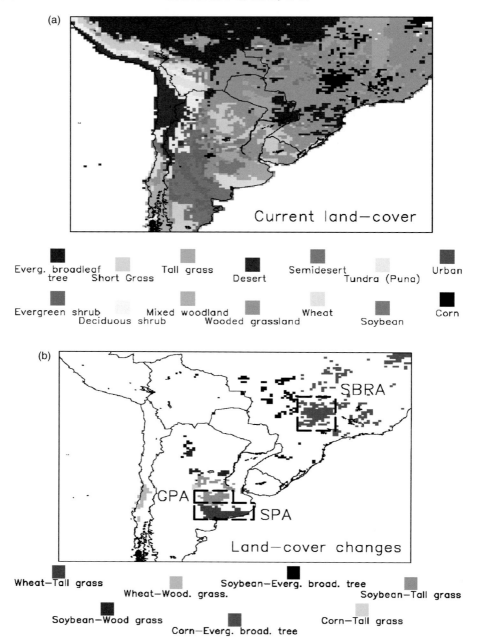

Figure 6.1 (a) Land cover for the Beltrán (2005) South America simulation domain. (b) Land-cover changes in the FS methodology RAMS application. CPA: central Pampas, SPA: southern Pampas, SBRA: southern Brazil areas mentioned in the text. See plate section for color version.

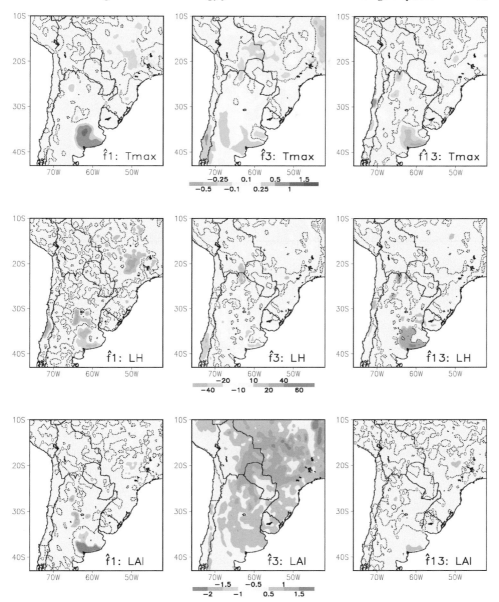

Figure 6.2 Contributions from natural vegetation, $\hat{f}1$ (left); $2 \times CO_2$ biology, $\hat{f}3$ (middle); and both, $\hat{f}13$ (right), for maximum temperature Tmax ($^\circ$C) (top row), diurnal latent heat flux LH (W m^{-2}) (middle row) and leaf area index LAI (m^2 m^{-2}) (bottom row) for the southern South American experiments. See plate section for color version.

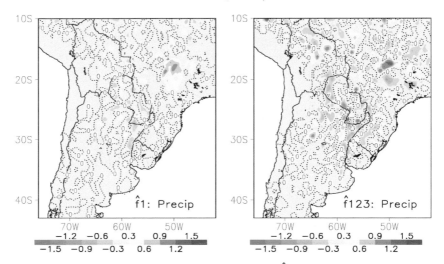

Figure 6.3 Contributions from natural vegetation, $\hat{f}1$ (left) and all natural vegetation, $2 \times CO_2$ biology and $2 \times CO_2$ radiation, $\hat{f}3$ (right) for daily precipitation
(mm day^{-1}) for the southern South American experiments. See plate section for
color version.

6.3.2 *Sensitivity to future LULC changes using projected future atmospheric scenarios*

Pitman and Narisma (2005) and Narisma and Pitman (2006) addressed the impacts
of future LULC, namely reforestation, on the Australian climate. They focused
on the potential of reforestation to mitigate the projected summer warming over
Australia under two emission scenarios for 2050 and 2100. They performed GEM-
RAMS ensemble simulations of the January climate over an Australian modeling
domain with a 56 km grid spacing. The atmospheric boundary conditions were
provided by the Commonwealth Scientific and Industrial Research Organization
(CSIRO) Mark 2 atmosphere–ocean model (Watterson and Dix, 2003). Each set
of LULC experiments used three types of land-cover scenarios: one current land-
cover or steady-state scenario (SS or f0 in the Alpert–Stein Methodology notation),
and two reforestation future scenarios (Table 6.4). In the low reforestation scenario
(LCL), 25% of the area that was converted from Eucalypt trees to grasslands in
the last 200 years was recovered. The high reforestation scenario (LCH) recovered
at least 75% of the deforested areas. The boundary conditions from the CSIRO
model are the 2050 and 2100 projected scenarios combined with high (A2) and
moderate (B2) CO_2 increase scenarios (i.e., 2050B2, 2050A2, 2100B2, 2100A2).
This example can be viewed as an application of the "fractional approach of the
methodology" (Krichak and Alpert, 2002), with only one varying factor (i.e., LULC

Table 6.4 *Summary of the experiments performed in Pitman and
Pitman (2005) and Narisma and Pitman (2006) for the low (LCL)
and high (LHC) reforestation scenarios, for each climate change
scenario (see text).*

	Land cover		
Experiment	Steady state	LCL	LHC
f0	✓		
f1_25		✓	
f1_75			✓

changes). The differences between the LCH, LCL, and SS land-cover experiments
represent the effects of the "pure" LULC contributions.

The simulations showed that reforestations have the potential to moderate the
future warming due to increases in CO_2 concentration in two of the three regions
of LULC changes. At a continental scale, the overall those impact is negligible,
but LULC change effects are very clear at a local scale and limited to the areas of
the LULC changes. The patterns of the impact are also directly associated with the
extent of the reforested area: the area of the impact is larger in the LCH case in both
2050 and 2100 scenarios than in the LCL case (Fig. 6.4). Fig. 6.4 also shows that the
cooling effect decreased in relative terms as the CO_2-induced warming intensified,
in the A2 scenario and for 2100. The analysis of the near-surface variables showed
that the cooling or warming effect of the future LULC changes is mainly due to
changes in the latent heat flux (i.e., increase and decrease respectively) related
to increases and reduced LAI values. The results of these simulations show that
although the impact of the LULC changes is at regional scales the effects remain
very important given that crops, water, and habitation exist at these scales.

6.4 Spatial pattern of land-cover change

Gero and Pitman (2006) used RAMS to assess the impact of land-cover change on a
simulated storm event over the Sydney basin in Australia. As urbanization expands,
the basin has increasingly transformed into a highly heterogeneous urban area, with
agriculture and natural vegetation coexisting within the basin. The summer storms
that frequently occur in the Sydney basin can be affected by the presence of the
urban surface, as suggested by many modeling and observational studies.

The initial experimental set-up used a high-resolution land-cover data set of
the Sydney basin's current land-use pattern and the natural land cover (i.e.,
pre-European settlement that occurred in 1778) to explore the potential impact

A. Beltrán-Przekurat, et al.

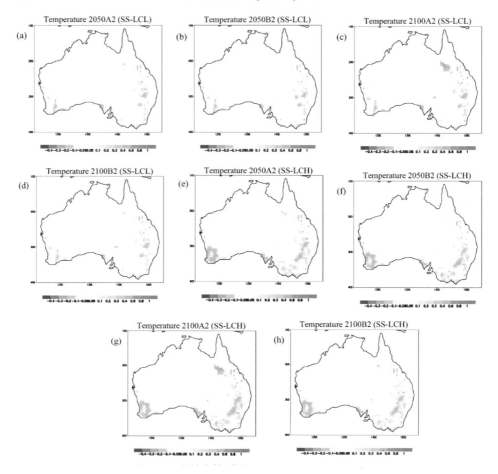

Figure 6.4 Differences in simulated surface air temperature (°C) between SS and LCL (left panels) in (a) 2050A2; (b) 2050B2; (c) 2100A2; (d) 2100B2; and between SS and LCH (right panels) in (e) 2050A2; (f) 2050B2; (g) 2100A2; (h) 2100B2. (From Narisma and Pitman, 2006) See plate section for color version.

of the LULC changes over the area. RAMS was configured with four nested grids, with horizontal increments of 60, 12, 3, and 1 km (Figure 6.5). The lateral boundary conditions were provided by the NCEP-NCAR reanalysis as in Eastman *et al.* (2001) to isolate the effects of LULC changes on a particular storm event. The natural land cover was assumed to be homogenous native bush, primarily 20 m Eucalypt forest that still exists in the area. The current land-cover types were generated from satellite images from Landsat 7 Enhanced Thematic Mapper Plus (Figure 6.5). The land-cover types included three specific urban areas within the Sydney basin (dense urban, new urban, established urban), agricultural land, and natural

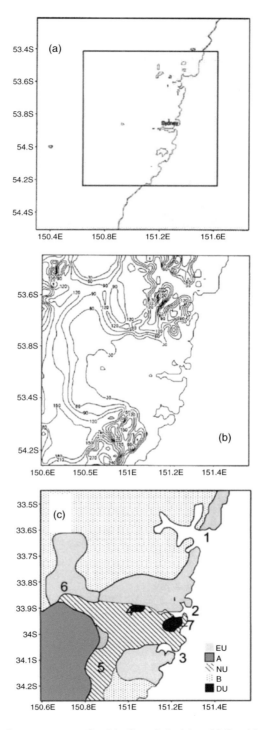

Figure 6.5 Nesting structure of grids 3 and 4: (a) grid 3 with inset of grid 4; (b) grid 4 with topography (meters above sea level); (c) current land cover for grid 4 (EU: established urban, A: agriculture, NU: new urban, B: bushland, DU: dense urban; 1: Broken Bay, 2: Port Jackson, 3: Botany Bay, 4: Parramatta, 5: Campbelltown, 6: Penrith, 7: Central Business District). (From Gero and Pitman, 2006)

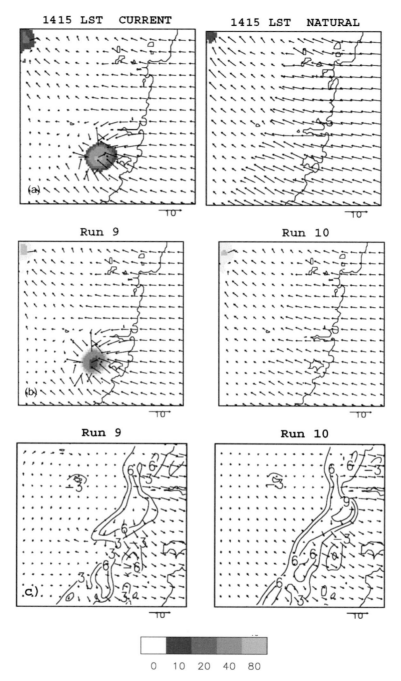

Figure 6.6 Simulated storm in grid 4 showing precipitation rate (gray shades; mm h^{-1}) and surface winds (arrowed vectors; m s^{-1}) at 1415 LST for (a) (left) current and (right) natural land-cover conditions and (b) (left) Run 9 and (right) Run 10 of the factorial simulations (see Table 6.5). (c) Horizontal divergence (10^4 s^{-1}) and winds at 850 hPa prior to the storm for (left) Run 9 and (right) Run 10 of the factorial simulations. (From Gero and Pitman, 2006)

Table 6.5 *Example of simulations for the factorial experiments in Gero and Pitman (2006). The areas of each type in the current land-cover experiment (agriculture, dense urban, new urban, established urban, and bushland (B)) are shown in Fig. 6.5. Simulations f0 and Run 9 generated the storm and f1 and Run 10 did not.*

	Agriculture		Dense urban		New urban		Established urban	
Exp	Yes	No (B)	Yes	No (B)	Yes	No (B)	Yes	No (B)
f0	✓		✓		✓		✓	
f1		✓		✓		✓		✓
Run 9	✓		✓			✓	✓	
Run 10		✓	✓			✓	✓	

vegetation (bushland). Each of them was associated with specific values in the biophysical parameters.

This initial set of experiments revealed that an intense convective storm developed in the model near Sydney's dense urban central business district under current land cover, but the storm was absent under natural land-cover conditions (Figure 6.6). To explore the effect of the dense and new urban surfaces on the generation of the storm under current land cover, an Alpert–Stein Methodology was implemented using the four current land-cover classes (three urban, and agriculture) as factors. All the possible combinations of land cover in the change from natural to current land cover, corresponding to 16 possible land-cover maps, were used to perform model simulations with the same NCEP-NCAR atmospheric boundary conditions. For example, Table 6.5 shows how the land cover in two of the runs of the factorial analysis was set up. In Run 9, the new urban area was returned to bushland, and in Run 10 both new urban and agricultural areas in the current land cover are returned to bushland. The factorial analysis revealed the storm to be sensitive to the presence of agricultural land in the southwest of the simulation domain (Figure 6.6). This area, through the roughness length, interacts with the sea breeze and affects the horizontal divergence: the smooth agricultural land exerts less drag on the atmosphere, allowing relative wind acceleration compared to the aerodynamically rough bushland.

6.5 Model/parameterization improvements versus better data

In this last example, RAMS is used to study the relative effects of incorporating a new model parameterization versus the use of improved data sets or boundary conditions. Lei *et al.* (2008) applied the Alpert–Stein Methodology to study the explicit

Table 6.6 *Summary of the experiments performed, the difference fields and their meaning in Lei et al. (2008)*

Exp	TEB		SST		Differences	Contributions from
	No	Yes	Clim	TRMM		
f0	✓		✓			
f1	✓			✓	$\hat{f}1 = f1 - f0$	Satellite SST data
f2		✓	✓		$\hat{f}2 = f2 - f0$	Urban model
f12		✓		✓	$\hat{f}12 = f12 - (f1 + f2) + f0$	Urban model and SST data

effects of an urban model and an improved sea-surface temperature (SST) data set, as well as their combined effect, on the simulation of a heavy rainfall event with RAMS. They studied a record-breaking event that occurred over Mumbai, India, on 25 July 2006, which was poorly represented by operational weather forecast models. Table 6.5 shows how the experiments were set up to study the relative impacts of improvements of the model (i.e., incorporation of an urban model) and the use of improved boundary conditions (i.e., observed SST).

Initially, experiments were carried out based on previous evidence that suggests that urban landscape can affect storm morphology. Lei *et al.* (2008) used RAMS coupled with TEB in the grid cells classified as urban in the innermost simulation domain, while LEAF2 was used as the default land-surface scheme in the rest of the grid cells (Figure 6.7). The experiments are f0 and f2 in Table 6.6. The urban heat island (UHI) created by the urban landscape helped to generate heavy precipitation over the Mumbai bay area, improving the simulated atmospheric conditions when RAMS without TEB was used.

Prior evidence also suggests that meso-scale boundaries can affect storm-related precipitation and the SST pattern is important for coastal city precipitation (Kumar *et al.*, 2007; Chang *et al.*, 2009). To explore the effects of SST as lower boundary conditions, a second set of simulations were performed, using the Alpert–Stein Methodology as the basis for the experiments. They used the SST fields from the Tropical Rainfall Measurement Mission (TRMM). Both the sole effects of SST without TEB ($\hat{f}1$ in Table 6.6) and the interactions between TEB and SST ($\hat{f}12$ in Table 6.6) were simulated. Through the Alpert–Stein Methodology Lei *et al.* (2008) found that the urban model itself was not the main reason for the better representation of rainfall fields (Figs. 6.7 and 6.8). The interactions between the UHI simulated by the urban model and the realistic satellite-derived SST contributed to accurately position and maintain the convergence zone and the observed heavy

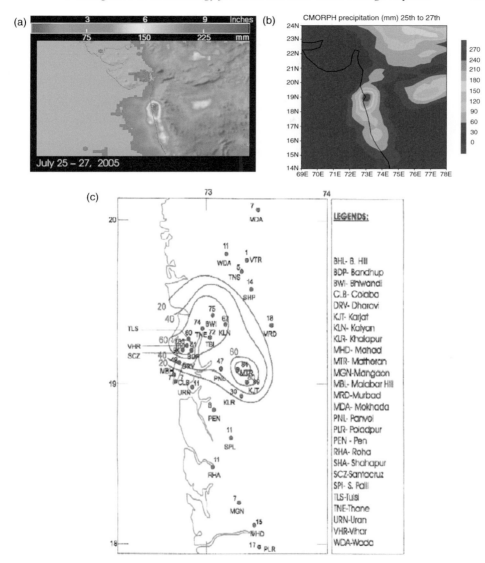

Figure 6.7 Total precipitation (mm) from different sources and from Alpert–Stein Methodology experiments: (a) TRMM; (b) CMORPH; (c) gauge data; (d) control, f0; (e) TRMM SST, f1; (f) Urban TEB, f2; (g) TRMM SST and Urban, f12. (From Lei *et al.*, 2008) See plate section for color version.

precipitation over Mumbai (Fig. 6.9). The local scale heterogeneities generated from the urban landscape and SST interactions, shown here in the convergence fields at 850 hPa (Fig. 6.8), were a crucial factor that improved the simulated rainfall.

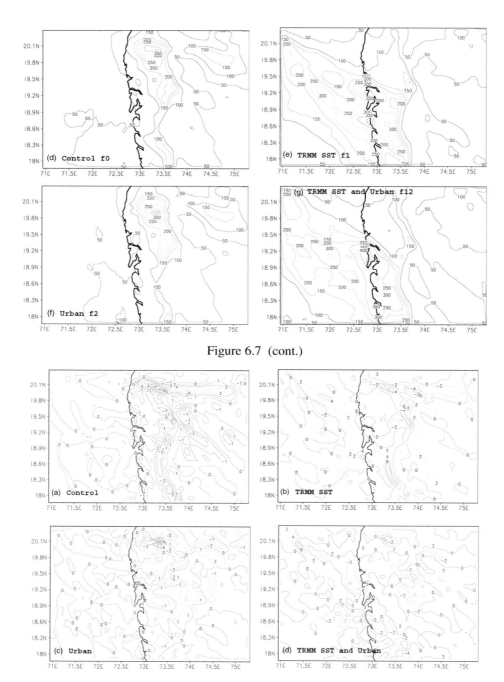

Figure 6.7 (cont.)

Figure 6.8 Factor separation values of single and interaction effects for the 850 hPa convergence (10^{-4} s^{-1}) on 09Z July 26: (a) control experiment, f0; (b) contribution from satellite SST data, $\hat{f}1$; (c) contribution from the urban model, $\hat{f}2$; (d) contribution from both the satellite SST data and the urban model, $\hat{f}12$. (From Lei *et al.*, 2008) See plate section for color version.

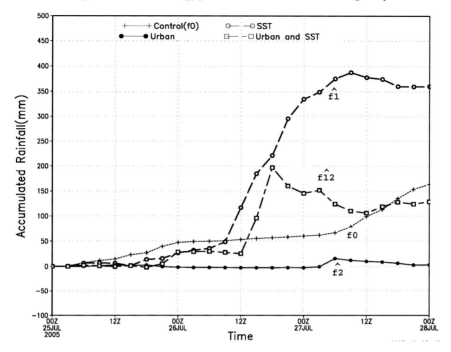

Figure 6.9 Time series of accumulated precipitation (mm) over Santacruz airport from different experiments: the control, the effect of TRMM SST, the effect of urban model, and the effect of their interactions. (From Lei *et al.*, 2008)

6.6 Final comments

Examples have been presented here of applications of the Alpert–Stein Factor Separation Methodology using RAMS, a regional mesoscale atmospheric model, for different regions in the world that had experienced important LULC, in addition to the changes in CO_2 levels. We also showed an example of future reforestation and climate scenarios, where the simulations involved the application of the fractional approach of the Alpert–Stein Methodology.

The Alpert–Stein Factor Separation Methodology, through these examples, allowed us to show the importance of including not only LCC scenarios in future climate change projection simulations but also the biological effects of CO_2. This indicates that regional and modeling systems need to include a land-surface scheme able to represent the processes of photosynthesis and soil and plant respiration, and the allocation of the assimilated carbon to leaves, stems, and roots.

Current and future LCC include increases of urban extension. Factorial experiments showed that a particular spatial configuration of a complex urban landscape initiated processes that led to the occurrence of a storm under current land cover. The presence of agricultural land in a specific area within the simulation domain and not the urban areas themselves triggered the storm. The last example of the application of Alpert–Stein Methodology showed that the interaction between urban parameterization and realistic satellite-derived SST improved the location of a storm and the simulation of the associated precipitation. The Alpert–Stein Methodology was able to identify the important role of the interaction between urban land cover and the meteorological processes. These results indicate the need to improve the parameterization of not only the particular urban surfaces but also of the other types of land cover at the periphery of the urban-complex areas. In addition, realistic representation of the current land cover may lead to more accurate weather predictions.

Acknowledgments

This work was supported in part by the Shortgrass Steppe Long Term Ecological Research project by funds from the NSF award DEB 0217631, NASA Grants NNX06AG74G and NNX07AG35G, NASA GWEC Grant NNG05GB41G, NOAA JCSDA Grant NA06NES4400013, DOE ARM 08ER64674, NSF Grants ATM-0296159 and ATM 0831331, NSF CAREER ATM-0847472. Roger A. Pielke Sr. was also supported in part by the University of Colorado at Boulder (CIRES/ATOC). Dallas Staley completed the editing of this paper in her usual outstanding manner.

7

Application of Factor Separation to heavy rainfall and cyclogenesis: Mediterranean examples

R. Romero

Heavy precipitation and intense cyclones exert a high societal impact in the Mediterranean region. Factor Separation (FS) by means of numerical simulations is a well-suited tool for improving our physical understanding of these phenomena. The roles played by *boundary factors*, such as the orography and surface heat fluxes, or by *physical factors*, such as the condensational latent heat release, have been isolated in numerous case studies. More recently, prognostic and diagnostic tools have been applied to also isolate the forcing associated with *dynamical factors*, such as the upper-level potential vorticity anomalies that tend to accompany the Mediterranean stormy weather.

7.1 Introduction

The Mediterranean basin is an ideal atmospheric research 'laboratory', recognized as one of the world's major cyclogenetic areas (Pettersen 1956, Radinovic, 1987). Much of the high-impact weather (notably strong winds and heavy precipitation) affecting its coastal countries has been statistically associated with the near presence of a distinct cyclonic signature (e.g., Jansá *et al.*, 2001). The numerical modelling of these atmospheric circulations is the most powerful tool available to scientists to develop a better physical understanding of the responsible mechanisms. In particular, many studies have used this potential to isolate the role played by different physical factors by means of the factor separation technique. *Boundary factors* (e.g., orography and latent heat flux from the Mediterranean) and *model physics factors* (e.g., latent heat release in cloud systems) have been typically considered. Some results on the role of both types of factors in Mediterranean flash flood events will be discussed in Section 7.2.

Factor Separation in the Atmosphere: Applications and Future Prospects, ed. Pinhas Alpert and Tatiana Sholokhman. Published by Cambridge University Press. © Cambridge University Press 2011.

Comparatively less attention, however, has been paid to the effects due to internal features of the flow dynamics (jet streaks, troughs, fronts, etc.) probably because, unlike the boundaries or model physics factors, modifying or switching off these elements in the simulations is not straightforward. The three-dimensional nature and mutual dependence of pressure, temperature and wind fields pose serious constraints on the ways these fields can be altered without compromising the delicate dynamical balances that govern both the model equations and actual atmospheric data. Section 7.3 presents a relatively clean approach to deal with these *dynamical factors*, based on the concept of potential vorticity (PV) and its invertibility principle (see Hoskins *et al.* (1985) and references therein, for an extended description of these important terms[1]). The role of upper-level precursor disturbances in a heavy rain producing western Mediterranean cyclone will be studied using this *PV inversion method*. The PV factor separation from the topography factor applying the Alpert–Stein Methodology was earlier performed by Tsidulko and Alpert (2001).

The applicability of the FS method to the study of extratropical cyclones in a framework that does not involve numerical model simulations will be highlighted in Section 7.4. Specifically, an experimental design that implements quantitatively the *PV thinking* concepts will be presented. It is based on a prognostic system of balance equations that are consistent with the PV inversion method. By switching on and off the PV anomalies of interest, different flow configurations are generated and the corresponding solutions to the prognostic equations can be algebraically combined to isolate the magnitude of both the individual and synergistic effects of the PV anomalies on the spatial pattern of geopotential height tendency – and vertical motion – around the cyclone, with the additional advantage of low computational cost. The potential of this novel approach to elucidate the impacts and interactions of the undulating tropopause, the low-level baroclinicity, and the latent heat release on a deep Mediterranean cyclone will be discussed in the section.

7.2 The role of boundary and model physics factors

The feasibility of mesoscale numerical modelling for the heavy rain problem in the Mediterranean has long been reported through numerous simulations of flash flood events. Some of these numerical studies have applied the FS method, a unique strategy to develop improved understanding of the responsible physical mechanisms. In this way, pre-existing conceptual models of the synoptic-mesoscale scenarios conducive to heavy precipitation have been confirmed – or rejected, depending on

---

[1] In short, the invertibility principle states that positive (negative) PV anomalies, or surrogate warm (cold) surface potential temperature anomalies, induce characteristic cyclonic (anticyclonic) flow patterns. Another important property of potential vorticity is its conservation under adiabatic and frictionless conditions (i.e., PV behaves as a tracer of atmospheric motion; the conservation principle).

the case – by isolating the explicit role of the studied factors, often related to surface characteristics of the model or to parameters/calculations of its physical parameterizations. These kinds of factors will be easily discernible by any advanced user of the model and are clearly susceptible to being switched on/off in the simulations. Two illustrative examples corresponding to flash flood situations that affected eastern Spain are here presented.

7.2.1 Orography and evaporation from the sea: 9–10 October 1994 event

The conceptual model of heavy precipitations in eastern Spain attributes a crucial role to the regional topography and the latent heat flux from the Mediterranean waters. Indeed, the topography of the western Mediterranean basin is very complex. At the local scale, it assists the development of clouds at fixed zones or in enhancing precipitation from pre-existing cloud systems, leading to accentuated rainfall differences between uplands and lowlands, or between slopes with different exposures to the maritime winds. At larger scales, it acts to transform the atmospheric disturbances that approach from the Atlantic, frequently generating secondary cyclones over the Mediterranean, mainly in the lee of the Atlas, Pyrenees and Alps (Reiter, 1975). With regard to eastern Spain, a shallow lee cyclone develops off the Algerian coast when the southwesterly flow of the downstream sector of the upper-level trough or cut-off low crosses the Atlas ridge almost perpendicularly. This mesoscale cyclone enhances or redirects the general easterly flow at low levels, thus promoting moisture flux convergence and focusing rainfall in the favorably exposed coastal areas. On the other hand, the closed characteristic of the Mediterranean sea and the high insolation received during the summer lead to high sea surface temperatures during the summer and autumn. This aspect ensures strong water vapour availability in the Mediterranean environment, and during that period the Mediterranean air masses frequently present convective instability (Meteorological Office, 1962; Ramis, 1995). This is premonitory of the torrential character of rainfall when, under favourable synoptic conditions, that instability is released over the coastal areas, sometimes causing local flooding.

The 9–10 October 1994 heavy precipitation event affected the eastern and northeastern regions of Spain, from south Valencia to Catalonia. In the latter region, up to 450 mm was registered at some locations, and the resulting flood can be considered one of the most devastating in recent decades. The simulations presented (taken from Ramis *et al.*, 1998) correspond to the second half of this event.

The rainfall spatial distribution is well captured by a control simulation (Fig. 7.1). A coastal band of more than 40 mm, resembling the observed pattern (not shown), is forecast by the model from central Valencia to northern Catalonia. Although rainfall

Figure 7.1 Forecast total precipitation from 06 UTC 10 October 1994 to 06 UTC 11 October 1994. Contour interval is 20 mm starting at 20 mm (continuous line). Dashed contour represents 10 mm. (Figure from Ramis *et al.*, 1998)

is underestimated, a large area exceeding 60 mm is obtained in Catalonia (peak values exceed 100 mm). Factor separation was conducted on this event after running three additional simulations with/without orography and sea evaporation in order to test the previous conceptual model. When considering the basic simulation f_0 (i.e., without both factors), less than 10 mm are given by the model in all the area of interest, including sea zones. This means that the considered factors become fundamental for the rainfall distribution and amounts. The isolated effects allow us to draw the following conclusions, in favour of the conceptual model but with unprecedented quantitative detail and putting much emphasis on the synergism of the factors:

• The orography is responsible for the development of the shallow low in the lee of the Atlas Mountains over the Algerian coast, as well as for the pressure dipole about the Pyrenees (Fig. 7.2a). These structures, which are more noticeable when the upper level flow is strong, lead to an intensification of the pressure gradient over the western Mediterranean and therefore to an enhancement of the moist easterly flux towards the Spanish coasts. In addition, the orography is also responsible for the warm air anomaly present at low levels over the south of the western Mediterranean. This warm air tongue and the easterly flux combine to reinforce the warm advection towards the Spanish coastal orography. The resulting contribution to the accumulated rainfall is very important (Fig. 7.2b): the

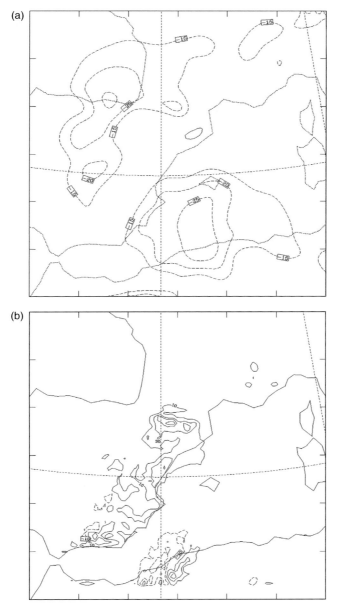

Figure 7.2 Effect of the orography: (a) on the geopotential field at 1000 hPa (in gpm) on 18 UTC 10 October 1994; (b) on the total precipitation at the end of the simulation (06 UTC 11 October 1994) (contour interval is 20 mm starting at 10 mm, continuous line, and at −10 mm, dashed line). (Figure from Ramis *et al.*, 1998)

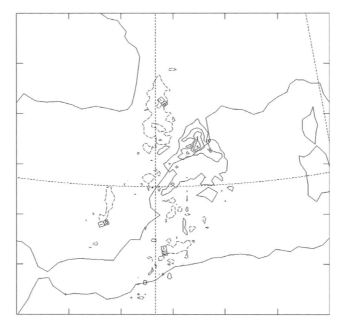

Figure 7.3 Effect of the interaction orography/evaporation on the total precipitation at the end of the simulation (06 UTC 11 October 1994) (contour interval is
20 mm starting at 10 mm, continuous line, and at −10 mm, dashed line). (Figure
from Ramis *et al.*, 1998)

individual orographic effect practically explains the forecast rainfall in all zones except
in coastal Catalonia and over the sea (compare with Fig. 7.1).

- The evaporation from the sea does not produce, when orography is not present, any
  remarkable effect on the precipitation, geopotential and wind fields (contribution not
  shown).
- The combination orography/evaporation was the most decisive factor for the spatial distribution of rainfall in the 9–10 October 1994 flash flood event. In particular, it is responsible
  for the coastal maxima located in Catalonia (Fig. 7.3). For other events (e.g., Romero
  et al., 1997, 1998), the synergistic effect was not so relevant because the air mass advected
  towards the coastal areas was already charged with enough moisture. However, the present case illustrates the great importance that latent heat fluxes during the course of the simulation may have in some situations, and suggests that quantitative precipitation forecasts
  in the western Mediterranean area may be especially sensitive to sea surface temperature.

7.2.2 Remote orography and latent heat release:
3–4 November 1987 event

This case was characterized by a quasistationary mesoscale convective system
(MCS) that developed over the Valencia region and lasted more than 30 h. Rainfall

totals exceeded 800 mm at some localities, producing one of the most catastrophic flash floods in the recent history of the region. The event occurred within a weak and very stagnant synoptic pattern under a persistent easterly-northeasterly low-level jet stream (LLJS) impinging on the Valencian orography.

Forecast precipitation for this event (from Romero *et al.*, 2000) is shown in Fig. 7.4a. The model is able to forecast quite accurately the location of the heavy precipitation centre in the Valencia region, but underestimates the peak amount (a forecast maximum of 231 mm vs. observations in excess of 600 mm). The model provides good guidance about the stationarity (measured as the amount of time with uninterrupted rainfall for each grid point) of the convective system that developed over Valencia (nearly 30 h; Fig. 7.4a).

The ability of the model to capture the location, long life and stationarity of the precipitation system seems to be associated with the correct prediction of the large-scale aspects of the circulation and its slow movement. However, the spatial localization and high precipitation efficiency of the system have to be related to the mesoscale ingredients. Indeed, the circulation pattern at low levels is rich in mesoscale components that can plausibly explain the continuous convective development over central Valencia (Fig. 7.4b). First of all, the low-pressure area extending leeward of the Atlas Mountains over the Mediterranean, with its centre over the Algerian coast, is contributing an increase of the easterly flow over Valencia. The simulation shows that this secondary low is maintained over the same area, probably owing to the constant interaction of the upper-level flow with the Atlas orography (recall the conceptual model). Embedded within the general low-pressure area of the southern Mediterranean, there is a mesolow over Valencia, approximately where the heaviest precipitation occurred. The genesis and maintenance of the mesolow is contemporaneous with the occurrence of precipitation over the Gulf of Valencia. This suggests that the mesolow could be the result of intense latent heat release in the atmospheric column. This hypothesis is further supported by the observation of a mid-tropospheric warm core in standard radiosonde measurements (not shown). The mesolow, stimulated by vigorous condensation over central Valencia, is associated with strong low-level convergence and therefore helps to maintain, or enhance, condensation and precipitation, a positive feedback loop. It seems from Fig. 7.4b that the mesolow is helping to intensify and concentrate the LLJS in its leading part. Once convection is initiated, the LLJS focuses the flow into the sloping terrain and produces intense low-level convergence over central Valencia (Fig. 7.4b) that persists throughout the episode.

Results of the previous experiment suggest that the effects of Atlas orography and latent heat exchange could have been quite relevant for this case study. This hypothesis can be investigated in a rigorous and quantitative way by means of FS. Consider first one simulation where both factors are eliminated. That is, the north

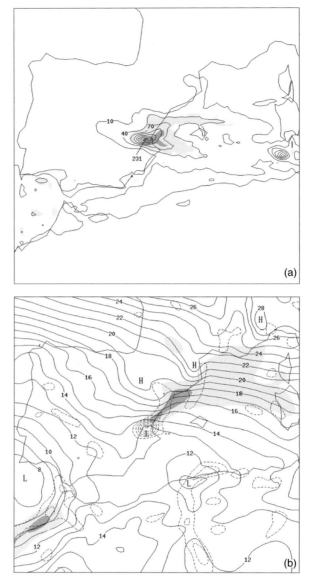

Figure 7.4 (a) Forecast total precipitation from 00 UTC 3 November to 12 UTC
4 November 1987. Contour interval is 30 mm, starting at 10 mm (maximum is
231 mm). Areas with continuous precipitation lasting more than 18 h and 27 h
are shown in light and dark shading respectively. (b) Model predicted sea-level
pressure (continuous line, in hPa without the leading 10 or 100), horizontal wind
convergence at 925 hPa (dashed line, contour interval is $5 \times 10^{-5} \, \text{s}^{-1}$, starting at
$5 \times 10^{-5} \, \text{s}^{-1}$), and horizontal wind speed at 925 hPa greater than 15 and 20 m s^{-1}
(light and dark shading respectively), for 18 UTC 3 November 1987. Lows and
highs mentioned in the text are indicated. (Figure from Romero *et al.* 2000)

African orography (a boundary factor) is removed prior to the initialization of the model, and latent heating terms (both absorption and release by the water phases; a model physics factor) appearing in the temperature tendency equation are explicitly set to zero. The results differ appreciably from the control or full experiment: there is a general suppression of rainfall in central Valencia and the complete disappearance of the inland peak, and the low-level circulation is appreciably modified, with much less convergence and a shift of the LLJS northwards (not shown). These effects confirm the crucial role of the factors, acting either individually or synergistically. The three possible contributions are isolated by means of the FS technique and are displayed in Fig. 7.5.

The modulation of the low-level circulation over the Mediterranean induced by the Atlas ridge is shown to be important. This modulation, caused by the cyclogenetic action leeward of the ridge, determines an enhancement of the easterly flow and horizontal convergence towards central Valencia, supposing a localization of

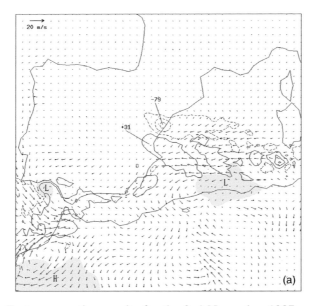

Figure 7.5 Factor separation results for the 3–4 November 1987 event. Effects of (a) Atlas orography, (b) latent heat exchange and (c) interaction of Atlas orography/latent heat exchange, on the fields: total precipitation, for which contour interval is 30 mm, starting at 10 mm for positive contributions (continuous line), and at −10 mm for negative contributions (dashed line) (some maxima are also indicated); vector wind field at 925 hPa on 21 UTC 3 November 1987 (reference vector is shown in the upper left corner); and sea-level pressure on 21 UTC 3 November 1987 (light shaded and dark shaded represent contributions greater than 0.5 and 2 hPa respectively, with L indicating zones of negative contribution and H zones of positive contribution). (Figure from Romero *et al.* 2000)

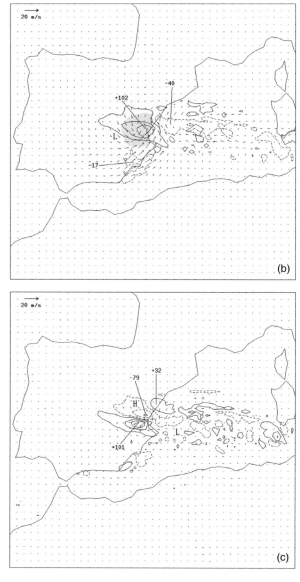

Figure 7.5 (cont.)

rainfall in that place at the expense of other areas further north (Fig. 7.5a). This result fits the classical conceptual model of flash flood situations in eastern Spain.

   The effects of atmospheric diabatic heating due to latent heat release reveal a strongly positive contribution of this factor (Fig. 7.5b). The mesolow noted earlier in the areas of heavy precipitation is entirely due to this factor. The associated cyclonic circulation affects the parent circulation strongly, enhancing the horizontal convergence ahead of the LLJS. Since this structure is continuously redeveloped in fixed areas of Valencia owing to the stationarity of the precipitation system, it results in strong rainfall localization in inland areas. These results reinforce the notion that the forecast of flooding situations requires not only a proper identification of ingredients prior to the development of convection, but also a close tracking and monitoring of the event after the onset of precipitation, trying to identify new elements that could endow the convective system with additional organization, efficiency or longevity.

Finally, the interaction of the Atlas orography and latent heat factors defines a rather complex pattern, not easily interpretable (Fig. 7.5c). Nevertheless, the effect on the rainfall field over Valencia seems to follow a similar tendency to the Atlas Mountains effect; that is, a southward shift of the rainfall activity. This implies a strengthening and displacement to the south (nearly at the location of the forecast maximum) of the rainfall induced by the latent heat factor.

7.3 Dealing with dynamical factors: the PV inversion method

On some occasions an in-depth understanding of the physical mechanisms responsible for a certain meteorological event will demand consideration of one or several internal features of the flow dynamics as factors. This is readily illustrated through the following example of a heavy-rain-producing western Mediterranean cyclone that impinged on eastern Spain on 28–29 September 1994, taken from Romero (2001).

7.3.1 Motivation: control simulation of the 28–29 September 1994 event

A two-way nested control simulation of this event (referred to as *S*) is summarized in a single composite chart for both the synoptic scale and mesoscale domains (Fig. 7.6). As shown in Fig. 7.6a, the upper-level flow at the beginning of the event was characterized by a positively tilted long-wave trough with its axis crossing the Iberian Peninsula. The induced flow over the western Mediterranean area at upper levels was thus from the southwest. Embedded within the large-scale trough, there are two geopotential height minima centred near the Gulf of Vizcaya and the Moroccan Atlantic coast. Calculation of Ertel's potential vorticity (Rossby, 1940; Ertel, 1942), defined as

$$q = \frac{1}{\rho}\eta \cdot \nabla\theta, \tag{7.1}$$

where ρ is the density, η the absolute vorticity vector, and θ the potential temperature, reveals that the two embedded lows are both associated with a PV maximum,

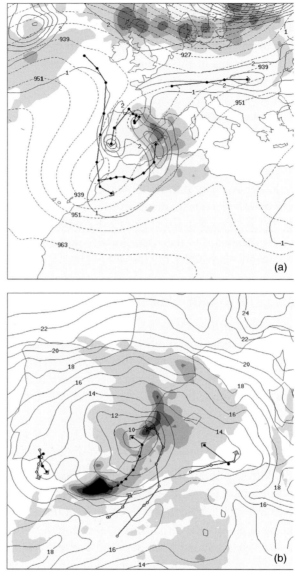

Figure 7.6 Control simulation *S*. (a) Geopotential height at 300 hPa at 06 UTC 28 September 1994 (dashed line, in dam); Ertel's potential vorticity on the 330 K isentropic surface at 00 UTC 30 September 1994 (continuous line, in PVU), with an indication of trajectories at 6 h intervals of several maxima present at that time (circles); and maximum value of the low tropospheric (1000–700 hPa) temperature advection observed during the simulation (shaded, for values exceeding 0.2, 0.4 and 0.6 K h^{-1}). (b) Sea-level pressure at 00 UTC 30 September 1994 (continuous line, in hPa without the leading 10), with an indication of back trajectories at 3 h intervals of several minima (circles, with open circles for minima dissipated before that time); and total precipitation (shaded, for values exceeding 10, 40, 70, 100 and 130 mm). (Figure from Romero, 2001)

as expected.[2] Figure 7.6a displays the trajectories of these PV centres on the 330 K isentropic surface during the simulation, as well as their structure and position at the end of the period. Since both diffusion and diabatic effects are active in the model, neither the PV nor the potential temperature are conserved, but the use of isentropic – and even isobaric – PV charts aloft, where the above processes are relatively weak, allows an easy identification and tracking of any PV structure. Observe that, while rotating about each other, the PV centres initially located in the Gulf of Vizcaya and the Moroccan coast migrate toward southern Portugal and the western Mediterranean, respectively. This movement implies high values of PV advection at upper levels over the western Mediterranean and eastern part of Spain progressing from south to north (not shown). The overall cyclonic rotation and lifting of the upper-level trough is also reflected in the trajectories of other secondary PV structures, some of which are included in Fig. 7.6a.

Associated with the evolution at upper levels, warm air advection in the lower troposphere during the simulation is maximized over the western Mediterranean and eastern Spain (see shaded field in Fig. 7.6a), again with the maximum of this field progressing from south to north. Additional diagnostic variables (not shown) help to describe the dynamics of this case: calculation of quasigeostrophic forcing for vertical motion reveals that the evolution of the upper-level PV field and the low-level temperature advection field are associated, respectively, with important centres of dynamic forcing for upward motion in the mid-upper and lower troposphere. These areas of positive forcing tend to overlap over the western Mediterranean and eastern Spain, and the overlapping zone moves northwards during 28–29 September, approximately following a cyclonic path along the Spanish Mediterranean coast. The model-predicted vertical velocity field nearly replicates the structure found for the forcing, although the magnitude of the upward motion at low levels along the eastern coast of Spain is increasingly dominated by the orographic action (not included in the quasigeostrophic formulation). In addition to a vertically coherent dynamic forcing field, the model forecasts large values of water vapour flux convergence at low levels in the same areas. Since convective instability, characteristic of the Mediterranean air mass during the warm season (Meteorological Office, 1962; Ramis, 1995), was also present, the synoptic environment was highly supportive for the development and maintenance of convection (Doswell *et al.*, 1996).

In response to the dynamical forcing pattern described previously, an intense cyclonic development and heavy precipitation are forecast over the western Mediterranean and eastern Iberian peninsula (Fig. 7.6b). Back trajectories for

---

[2] The reader should note that positive/negative PV anomalies will be found in connection with cyclonic/anticyclonic vorticity signals of the atmospheric circulation such as the very familiar jet streaks, troughs, ridges, cut-off lows, fronts, etc. displayed in routine meteorological charts.

several sea-level pressure minima identified during the simulation have been incorporated into the figure to illustrate the evolution of the surface disturbance. It is interesting to note that, unlike many cases of flash flood events in eastern Spain, characterized by relatively stationary surface lows near the Algerian coast (recall the cases from the last section), in this case the disturbance is mobile and progresses northwards from that genesis area to the coast of Valencia. This behaviour appears to be associated with the strong and evolving dynamic forcing identified aloft.

Although the two embedded upper-level PV centres identified on Fig. 7.6a seem to be playing an important role in the evolution, intensity and areal extent of the surface cyclone, there is no practical way, with a single control simulation, of quantifying the degree of dependence of the mesoscale forecast on the specific structure of the upper-level flow. Additional numerical simulations with the upper-level PV centres playing the role of explicit factors appear to be necessary. The object is to ascertain if the development and track of the surface cyclone, as well as its associated rainfall, are sensitive to the intensity and position of the two embedded PV centres. It must be noted that these relatively small-scale components of the flow may be subject to significant uncertainty as a consequence of analysis and/or model errors. Here we will take advantage of the PV invertibility principle.

7.3.2 The PV inversion method

According to the invertibility principle, given some balanced flow constraints and proper boundary conditions for the meteorological fields (pressure, temperature and wind), the knowledge of the three-dimensional distribution of PV can be used to infer the balanced meteorological fields. Application of the so-called piecewise PV inversion is particularly useful, since once identified, any PV element of interest as well as its associated mass and wind fields can be isolated in a consistent way for diagnostic or prognostic purposes (i.e., the PV element becomes a tangible factor). The piecewise PV inversion technique of Davis and Emanuel (1991) will be used to explore the sensitivity of the 28–29 September 1994 mesoscale simulation to the upper-level PV centres, for convenience hereinafter referred to as SW and NE PV centres according to their initial geographical location (Fig. 7.6a).

The method starts with the calculation of the balanced flow, described by ϕ (geopotential) and ψ (streamfunction), from the total or instantaneous distribution of Ertel's potential vorticity q, given by expression (7.1). The balance assumption made herein follows the Charney (1955) nonlinear balance equation:

$$\nabla^2 \phi = \nabla \cdot f \nabla \psi + 2m^2 \left[\frac{\partial^2 \psi}{\partial x^2} \frac{\partial^2 \psi}{\partial y^2} - \left(\frac{\partial^2 \psi}{\partial x \partial y} \right)^2 \right] \qquad (7.2)$$

where f is the Coriolis parameter and m denotes the map-scale factor of the particular (x, y) projection used. The other diagnostic relation necessary for the inversion of ϕ and ψ is given by the approximate form of Eq. (7.1) resulting from the hydrostatic assumption and the same scale analysis used to derive (7.2), namely, that the irrotational component of the wind is very small relative to the nondivergent wind:

$$q = \frac{g \kappa \pi}{p} \left[(f + m^2 \nabla^2 \psi) \frac{\partial^2 \phi}{\partial \pi^2} - m^2 \left(\frac{\partial^2 \psi}{\partial x \partial \pi} \frac{\partial^2 \phi}{\partial x \partial \pi} + \frac{\partial^2 \psi}{\partial y \partial \pi} \frac{\partial^2 \phi}{\partial y \partial \pi} \right) \right] \qquad (7.3)$$

where p is the pressure, g the gravity, $\kappa = R_d/C_p$ and the vertical coordinate π is the Exner function $C_p (p/p_0)^\kappa$.

The finite-difference form of the closed system described by (7.2) and (7.3) is solved for the unknowns ϕ and ψ given q, using an iterative technique until convergence of the solutions is simultaneously reached. Neumann-type conditions $(\partial \phi / \partial \pi = f \partial \psi / \partial \pi = -\theta$, where θ is potential temperature) are applied on the top and bottom boundaries, and Dirichlet conditions on the lateral boundaries. The latter are supplied by the observed geopotential and a streamfunction calculated by matching its gradient along the edge of each isobaric surface to the observed normal wind component, which is first slightly modified to force no net divergence in the domain. Because of the balance condition used, the inverted fields are very accurate even for meteorological systems characterized by large Rossby numbers.[3]

Next, a reference state must be found from which to define perturbations. This reference state is commonly defined as a time average. Given $\bar{q}$, the time mean of q, a balanced mean flow $(\bar{\phi}, \bar{\psi})$ is inverted from equations identical to (7.2) and (7.3), except all dependent variables are mean values and the mean potential temperature, $\bar{\theta}$, is used for the top and bottom boundary conditions. The perturbation fields (q', ϕ', ψ') are given by the definitions

$$(q, \phi, \psi) = (\bar{q}, \bar{\phi}, \bar{\psi}) + (q', \phi', \psi') \qquad (7.4)$$

Finally, we can consider that the PV perturbation field q' is partitioned into N portions or anomalies:

$$q' = \sum_{n=1}^{N} q_n \qquad (7.5)$$

---

[3] The more traditional quasigeostrophic formulation of PV inversion becomes simpler to implement than this nonlinear method, but at the price of its results being accurate for small Rossby numbers only.

We are interested in obtaining that part of the flow (ϕ_n, ψ_n) associated with each PV portion q_n, and we also require

$$\phi' = \sum_{n=1}^{N} \phi_n, \qquad \psi' = \sum_{n=1}^{N} \psi_n \qquad (7.6)$$

As discussed in Davis (1992), there is no unique way to define a relationship between (ϕ_n, ψ_n) and q_n because of the nonlinearities present in Eqs. (7.2) and (7.3). We adopt the linear method of Davis and Emanuel (1991), derived after substitution of expression (7.4) and the above summations in Eqs. (7.2) and (7.3) and equal partitioning of the nonlinear term between the other two linear terms that result from each nonlinearity in the above equations. The resulting linear closed system for the nth perturbation is

$$\nabla^2 \phi_n = \nabla \cdot f \nabla \psi_n + 2m^2 \left(\frac{\partial^2 \psi^*}{\partial x^2} \frac{\partial^2 \psi_n}{\partial y^2} + \frac{\partial^2 \psi^*}{\partial y^2} \frac{\partial^2 \psi_n}{\partial x^2} - 2 \frac{\partial^2 \psi^*}{\partial x \partial y} \frac{\partial^2 \psi_n}{\partial y \partial x} \right) \qquad (7.7)$$

$$q_n = \frac{g \kappa \pi}{p} \left[(f + m^2 \nabla^2 \psi^*) \frac{\partial^2 \phi_n}{\partial \pi^2} + m^2 \frac{\partial^2 \phi^*}{\partial \pi^2} \nabla^2 \psi_n \right.$$
$$\left. - m^2 \left(\frac{\partial^2 \phi^*}{\partial x \partial \pi} \frac{\partial^2 \psi_n}{\partial x \partial \pi} + \frac{\partial^2 \phi^*}{\partial y \partial \pi} \frac{\partial^2 \psi_n}{\partial y \partial \pi} \right) - m^2 \left(\frac{\partial^2 \psi^*}{\partial x \partial \pi} \frac{\partial^2 \phi_n}{\partial x \partial \pi} + \frac{\partial^2 \psi^*}{\partial y \partial \pi} \frac{\partial^2 \phi_n}{\partial y \partial \pi} \right) \right]$$

$$(7.8)$$

where $()^* = \overline{()} + \frac{1}{2}()'$.

The system (7.7)–(7.8) is solved for each PV anomaly of interest. A portion θ_n of the perturbation potential temperature can be used for the top and bottom boundary conditions, whereas homogeneous boundary conditions for ϕ_n and ψ_n are normally assumed at the lateral boundaries for interior PV anomalies such as the ones considered.

The piecewise PV inversion scheme has been applied to invert the SW and NE PV anomalies at 00 UTC 28 September 1994 – the simulation start time. The perturbation PV field is defined as the departure from the 6-day time average about that time. This mean state captures effectively the large-scale trough seen in Fig. 7.6a, i.e., the intrusion of a tongue of high PV towards the Iberian peninsula and lower latitudes (not shown). The pieces representing the two anomalies are identified as the volumes of positive PV perturbation above 500 hPa present to the southwest and northeast of the Gulf of Cádiz.

A vertical cross section of the inverted balanced fields (Fig. 7.7) illustrates that each localized PV anomaly is associated with geopotential deficit, cyclonic circulation, and cooling (warming) under (above) it. Although this response is stronger at

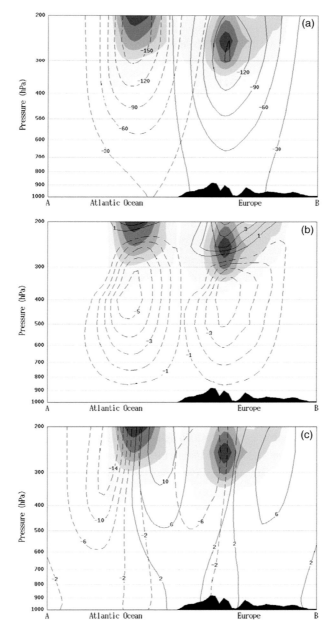

Figure 7.7 Vertical cross section along a SW-NE direction line crossing the Iberian Peninsula of the PV-inverted fields from the SW and NE anomalies at 00 UTC 28 September 1994. (a) Geopotential height (dashed and continuous contours for the SW and NE anomalies respectively, starting at −30 m every −30 m). (b) Temperature (continuous line for positive values and dashed line for negative values, in 1 °C intervals starting at 1 and −1 °C, respectively). (c) Section-normal winds (continuous line for flow into the page and dashed line for flow out of the page, in 4 m s^{-1} intervals starting at 2 and −2 m s^{-1}, respectively). The SW and NE PV anomalies are shown as shaded for values exceeding 0.5, 1.5, 2.5, 3.5 and 4.5 PVU. (Figure from Romero, 2001)

mid-upper levels, the effects are also felt in the lower troposphere, a natural consequence of the Laplacian operators involved in the system of equations (7.7)–(7.8).
This emphasizes that a potential error in resolving the upper-level PV centres would
be reflected not only at the anomaly level, but throughout the troposphere, down
to the surface level, affecting key fields for surface cyclogenesis and forcing of
vertical motion such as the temperature advection pattern.

7.3.3 *Role of the PV anomalies in comparison to the boundary factors*

Rather than performing a factor separation (e.g., aimed at calculating the synergy
between SW and NE PV anomalies on the cyclone evolution), in this academic
exercise we simply analyze, by means of independent numerical experiments,
the sensitivity of the forecast to the sub-synoptic structure of the upper-level
trough. These sensitivity experiments are designed by adding and/or subtracting
the PV-inverted balanced fields (geopotential, temperature and wind; Fig. 7.7) into
the model initial conditions,[4] very much what it is done in a classical FS exercise dealing with boundary and model physics factors. Two sets of simulations are
designed to study separately the sensitivity of the forecast to the intensity and position of the anomalies (Tables 7.1 and 7.2, respectively). In the first set, the SW and
NE anomalies are either doubled (adding the inverted fields), removed (subtracting
the fields) or kept unchanged, entailing eight simulations in addition to the control
run from all the possible combinations (summarized in Table 7.1). In the second
set, the intensity of the anomalies is not changed but the position is shifted along
the SW-NE direction. The anomalies are either moved outwards (by subtracting
the associated fields and adding them 425 km farther from the Iberian peninsula),
moved inwards (in the same way except 425 km closer to the Iberian peninsula) or
kept in the original position, implying again eight additional simulations (Table 7.2).

The above-mentioned ensemble of experiments is arbitrarily defined, without
regard to any PV error climatology as would be required for a potential application
of the method to ensemble forecasting. Nevertheless, the designed experiments
appear to embrace a wide range of synoptic scenarios which, while looking similar
from a broad perspective (since all share the presence of a well-defined trough
upstream from the Mediterranean, in agreement with the climatology of these flash
flood situations), contain differences that turn out to be crucial for the mesoscale
forecast.

Indeed, focusing on the gross features of the forecast flows, three main situations
are obtained. The *first group*, consisting of experiments with the anomalies removed

---

[4] The relative humidity field is kept unaltered in these experiments.

Table 7.1 *Summary of the numerical experiments performed in order to investigate the sensitivity of the 28–29 September 1994 simulation to the intensity of the upper-level PV anomalies*

Experiment	SW anomaly	NE anomaly
S_0^0	Removed	Removed
S_2^2	Doubled	Doubled
S_1^0	Unchanged	Removed
S_2^0	Doubled	Removed
S_0^1	Removed	Unchanged
S_0^2	Removed	Doubled
S_2^1	Doubled	Unchanged
S_1^2	Unchanged	Doubled

Table 7.2 *Summary of the numerical experiments performed in order to investigate the sensitivity of the 28–29 September 1994 simulation to the position of the upper-level PV anomalies*

Experiment	SW anomaly	NE anomaly
S_-^-	Moved inwards	Moved inwards
S_+^+	Moved outwards	Moved outwards
$S_=^-$	Unchanged	Moved inwards
S_+^-	Moved outwards	Moved inwards
$S_-^=$	Moved inwards	Unchanged
S_-^+	Moved inwards	Moved outwards
$S_+^=$	Moved outwards	Unchanged
$S_=^+$	Unchanged	Moved outwards

or moved away from each other (S_0^0, S_1^0, S_+^+ and $S_=^+$), results in synoptically weak scenarios, characterized by stationary surface lows extended along the lee of the Atlas Mountains and with most of the rainfall restricted to the southern Mediterranean areas (Fig. 7.8a). At the other end, a *second group* of experiments with enhanced PV structures aloft (S_2^2, S_-^-, $S_=^-$ and $S_-^=$) results in extensive and very mobile surface disturbances that generate heavy rain in the northern Mediterranean zones as well (Fig. 7.8b). And the *third group*, in which the relative weight of the northern anomaly is enhanced (S_0^1, S_0^2, S_1^2 and $S_+^=$), tends to produce cyclones that evolve farther east and north of southeastern Spain, thus inducing a concentration of most of the rainfall in northern Mediterranean areas (Fig. 7.8c). Of the remaining five experiments, S_2^0 produces a surface low and rainfall resembling those in the first group, and the other four members (S_2^1, S_+^-, S_-^+ and the control simulation S) would be classified in the second group according to its effects.

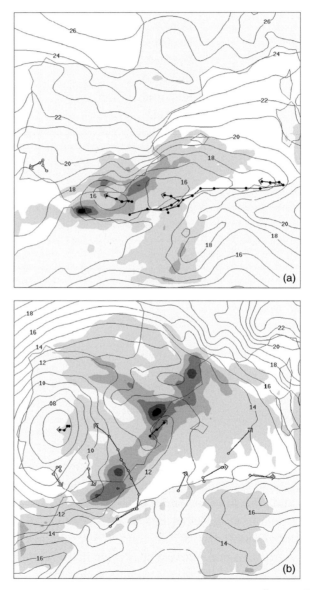

Figure 7.8 As in Fig. 7.6b, except for the experiments (a) S_0^0, (b) S_2^2 and (c) S_0^1. (Figure from Romero, 2001)

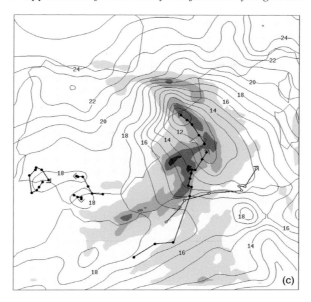

Figure 7.8 (cont.)

Finally, it would be interesting to judge the relevance of the upper-level PV anomalies relative to the action of other noninternal factors traditionally assumed – and also proved – to be very important in the western Mediterranean flash flood situations, notably the orography and the sea surface latent heat flux (recall last section). Note that for all the numerical experiments presented, there is a tendency for low-pressure developments in the lee of the Atlas Mountains, as well as precipitation enhancement in the exposed areas of eastern Spain. This suggests that both local and remote orographies could have played an important role in this case too. On the other hand, the simulations reveal intense evaporation from the warm ($\sim$23 °C) Mediterranean waters during the episode (not shown). For comparison, the control experiment S and the basic experiment S_0^0 were repeated, but eliminating the previous boundary factors (Fig. 7.9). The first experiment (Fig. 7.9a) still develops a large cyclone, but less intense and more circular than in the full simulation (compare with Fig. 7.6b). Clearly, the effect of the Atlas Mountains in modulating the sea-level pressure field over the Mediterranean is notable, which implies an enhancement of the impinging easterly moist flow. The general area of precipitation in this modified control experiment does not change strongly, but the amounts are reduced. The output of the modified S_0^0 experiment (Fig. 7.9b), to be compared with the results of Fig. 7.8a, does not even contain any noticeable low-pressure centre over the Mediterranean, and the precipitation produced is very weak. In conclusion, the external factors induced an appreciable modulation of the surface circulation and enhanced the efficiency of the system as a rainfall producer,

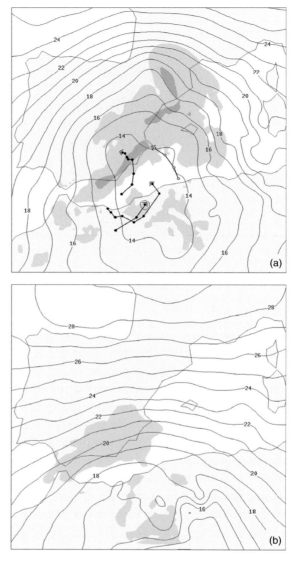

Figure 7.9 As in Fig. 7.6b, except for: (a) experiment S without both orography and sea surface latent heat flux; (b) experiment S_0^0 without both orography and sea surface latent heat flux. (Figure from Romero, 2001)

but the cyclogenesis that took place over the southern Mediterranean and its pro-gression to the north must be attributed mostly to the action of the upper-level PV anomalies. In a similar study, Tsidulko and Alpert (2001) isolated the synergism of upper-level PV and the Alps in a Genoa lee cyclogenesis event, illustrating that the PV/topography synergy is a leading contributor to lee cyclogenesis.

7.4 PV thinking of cyclones assisted by the Factor Separation method

Since the seminal paper of Hoskins *et al.* (1985), the use of potential vorticity to analyze the genesis and evolution of synoptic-scale systems has become very popular in meteorology. The two powerful principles of conservation and invertibility can be readily combined to develop a conceptually elegant framework that extends the capabilities of the more traditional quasigeostrophic theory for explaining the dynamics of mid-latitude circulation systems. This method of dynamical analysis has been referred to as 'PV thinking'.

Extratropical cyclogenesis is the best paradigm of cooperation between anomalies of different origin to reinforce each other. The vertical interaction between the upper-level wavelike PV anomaly and the potential temperature field along the lower boundary is fundamental to explain the baroclinic growth of the disturbance (Hoskins *et al.*, 1985). As the upper-level cyclonic PV anomaly arrives over a region of significant low-level baroclinicity, its induced circulation will promote warm advection east of it, creating a warm anomaly at the surface, which in turn will induce its own cyclonic circulation that is felt just to the east of the upper-level positive PV anomaly, thus contributing to its amplification by the southward advection of the high PV values found at higher latitudes; a positive feedback mechanism that will be reflected as the growth of the wave–cyclone system in this idealized dry atmosphere. For the real, moist atmosphere, the generation of low-level PV anomalies is due not only to advection but also to differential surface heating and condensation of water vapour in the atmospheric column. The role of latent heat release for accelerating the growth of the disturbance is well recognized, and this factor has been included explicitly in the PV framework by some authors (e.g., Davis and Emanuel, 1991).

In this section we present a 'dynamic' approach that allows us to isolate quantitatively the previous impacts and interactions of the PV anomalies during the life cycle of the cyclone. The method utilizes the FS technique, applied to the prognostic system of balance equations associated with the piecewise PV inversion technique described in Section 7.3. The potentialities of this method will be illustrated for the 10–12 November 2001 western Mediterranean cyclone, the worst storm affecting the Balearic Islands during recent decades (further details can be found in Romero, 2008).

7.4.1 The Mediterranean cyclone of 10–12 November 2001

This extraordinary cyclone originated from the north African lands over a region of marked baroclinicity, preceded by a significant cold air intrusion at upper levels from north-central Europe towards Iberia and Morocco. During this incipient phase on 9–10 November, catastrophic flash floods occurred in Algeria and Morocco. As the cyclone evolved northeastwards into the western Mediterranean basin, the central pressure continued to deepen and an appreciable pressure gradient developed around its core, leading to the mature or most intense state of the cyclone at around 11 November 00 UTC (Fig. 7.10a). At this time the warm and cold fronts' signatures in the low-level thermal field are both very clear and occlusion is suggested near the tip of the warm air surge from north Africa induced by the circulation (same figure). Meanwhile, the upper-level circulation adopted cut-off characteristics and two geopotential height minima are visible, located to the west and east of the

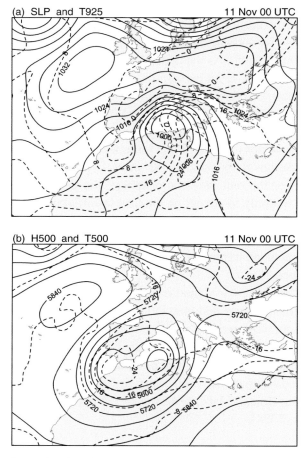

Figure 7.10 Synoptic situation on 11 November 2001 00 UTC. The following fields are shown: (a) sea-level pressure (solid line using 4 hPa contour intervals) and temperature at 925 hPa (dashed line using 4 °C intervals); (b) geopotential height at 500 hPa (solid line using 60 gpm contour intervals) and temperature at 500 hPa (dashed line using 4 °C intervals). (Figure from Romero, 2008)

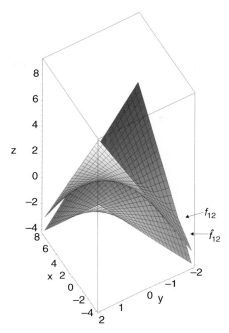

Figure 3.1 Graph for bivariate polynomial for functions f_{12}, $\hat{f}_{12}$, where x-axis and y-axis are x_1 and x_2, and z-axis is the field f. Upper functions f_{12} and lower, $\hat{f}_{12}$.

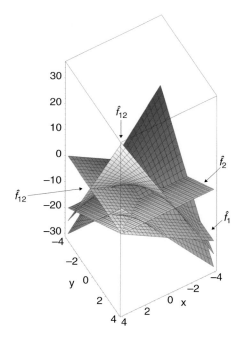

Figure 3.2 Graphs of functions: bivariate polynomial f_{12}, $\hat{f}_{12}$, $\hat{f}_1$, and $\hat{f}_2$.

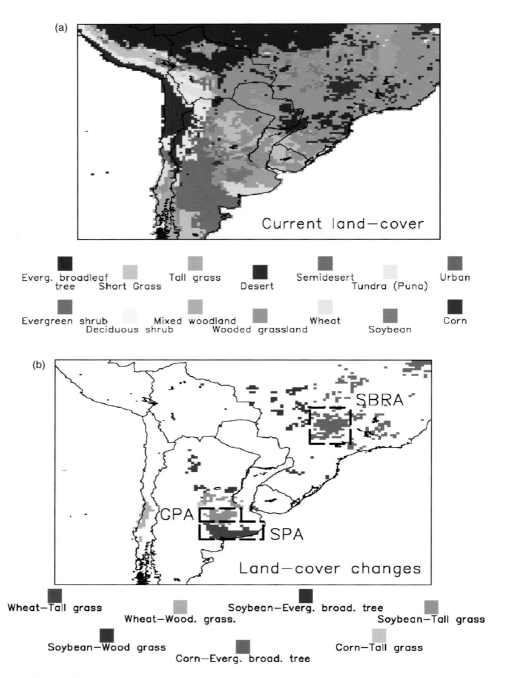

Figure 6.1 (a) Land cover for the Beltrán (2005) South America simulation domain. (b) Land-cover changes in the FS methodology RAMS application. CPA: central Pampas, SPA: southern Pampas, SBRA: southern Brazil areas mentioned in the text.

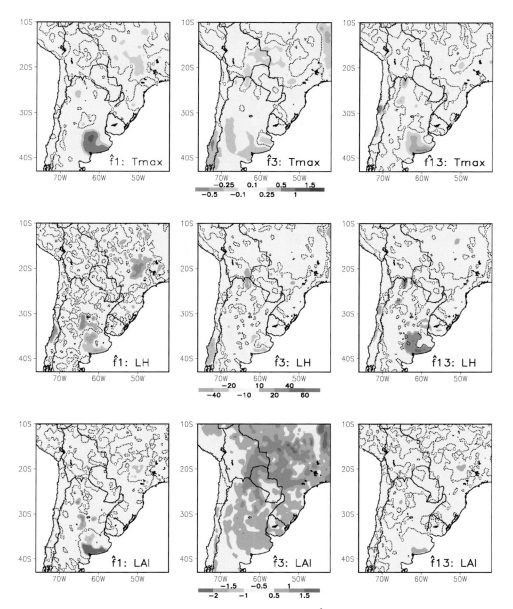

Figure 6.2 Contributions from natural vegetation, $\hat{f}1$ (left); $2 \times CO_2$ biology, $\hat{f}3$ (middle); and both, $\hat{f}13$ (right), for maximum temperature Tmax (°C) (top row), diurnal latent heat flux LH (W m^{-2}) (middle row) and leaf area index LAI (m^2 m^{-2}) (bottom row) for the southern South American experiments.

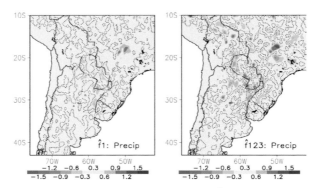

Figure 6.3 Contributions from natural vegetation, $\hat{f}1$ (left) and all natural vegetation, $2 \times CO_2$ biology and $2 \times CO_2$ radiation, $\hat{f}3$ (right) for daily precipitation (mm day^{-1}) for the southern South American experiments.

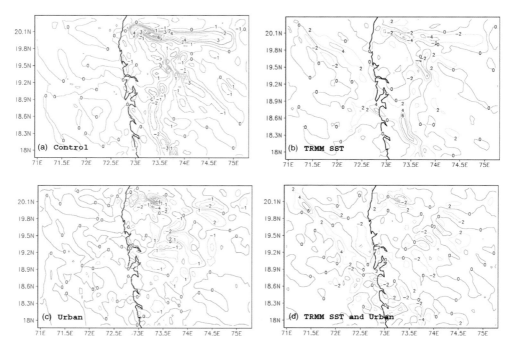

Figure 6.8 Factor separation values of single and interaction effects for the 850 hPa convergence ($10-4$ s^{-1}) on 09Z July 26: (a) control experiment, f0; (b) contribution from satellite SST data, $\hat{f}1$; (c) contribution from the urban model, $\hat{f}2$; (d) contribution from both the satellite SST data and the urban model, $\hat{f}12$. (From Lei *et al.*, 2008)

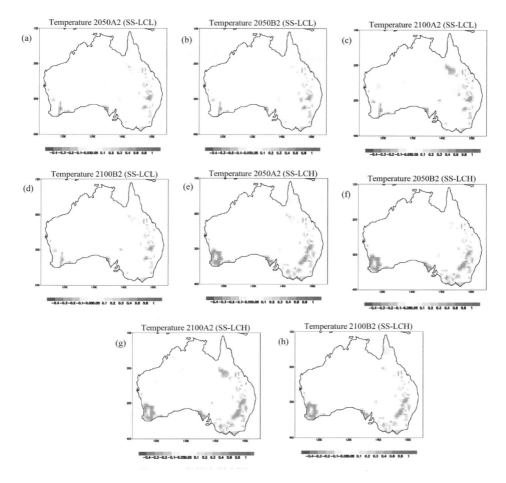

Figure 6.4 Differences in simulated surface air temperature (°C) between SS and LCL (left panels) in (a) 2050A2; (b) 2050B2; (c) 2100A2; (d) 2100B2; and between SS and LCH (right panels) in (e) 2050A2; (f) 2050B2; (g) 2100A2; (h) 2100B2. (From Narisma and Pitman, 2006)

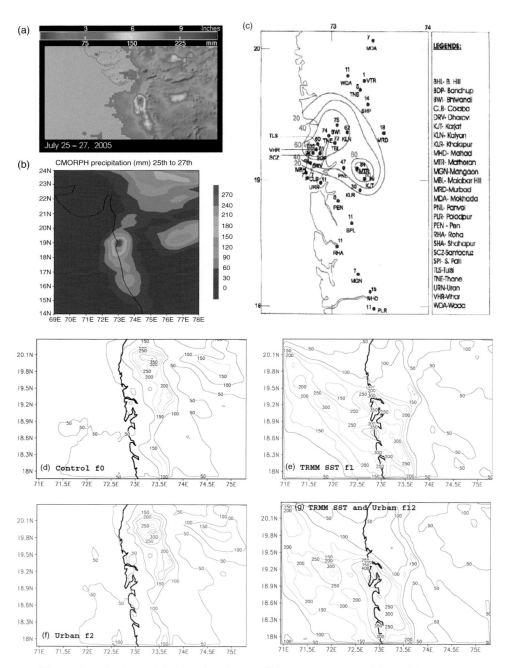

Figure 6.7 Total precipitation (mm) from different sources and from FS analysis experiments: (a) TRMM; (b) CMORPH; (c) gauge data; (d) control, f0; (e) TRMM SST, f1; (f) Urban TEB, f2; (g) TRMM SST and Urban, f12. (From Lei *et al.*, 2008)

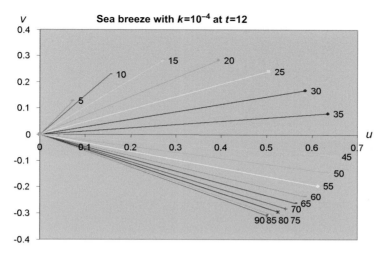

Figure 12.9 The breeze at $t = 12$ with friction coefficient $k_f = 10^{-4}\,\text{s}^{-1}$. The wind vector applied to different Coriolis parameters correspond to latitudes of $0°$ to $90°$. See plates section for color version.

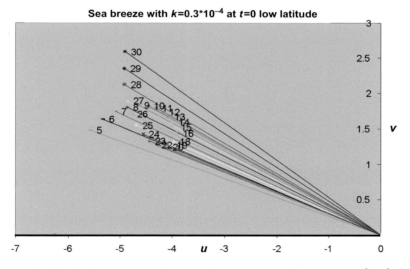

Figure 12.10 The breeze at $t = 0$ with a friction coefficient $0.3 \times 10^{-4}\,\text{s}^{-1}$. The wind vector applied to different Coriolis parameters corresponds to the latitudes of $0°$ to $30°$. See plates section for color version.

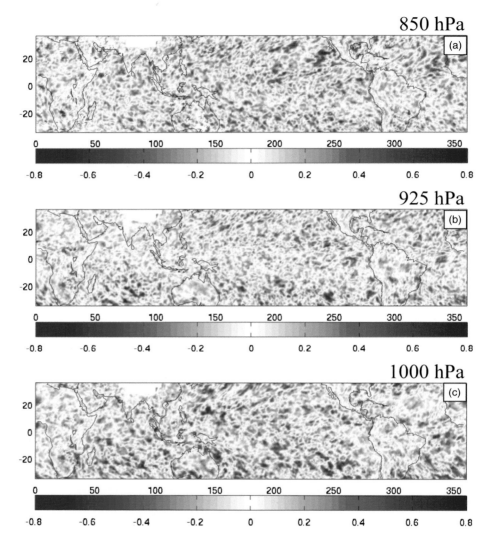

Figure 14.4 Spatial distribution of the weights (λ) calculated from multiple regression analysis with the 6 hr data from 23 March to 11 July 2007 at (a) 850 hPa, (b) 925 hPa, (c) 1000 hPa. Scale provided below each panel. Red shading denotes the positive values of weights while blue color denotes the negative values of weights.

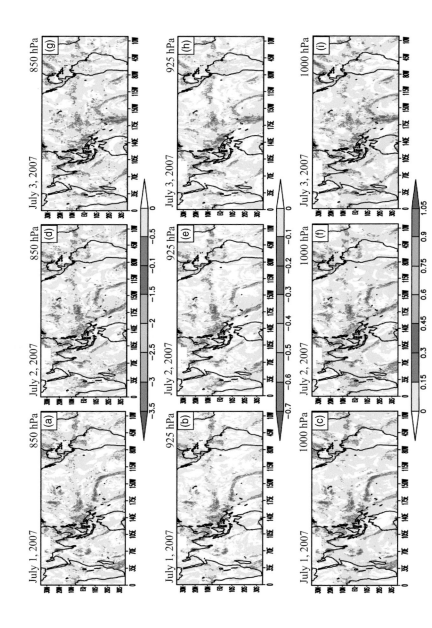

Figure 14.5 Diffusion term ($\times 10^{-4}$ s^{-1}) calculated from model forecasts at three vertical levels 850 hPa, 925 hPa, and 1000 hPa for (a)–(c) 1 July 2007, (d)–(f) 2 July 2007, and (g)–(i) 3 July 2007.

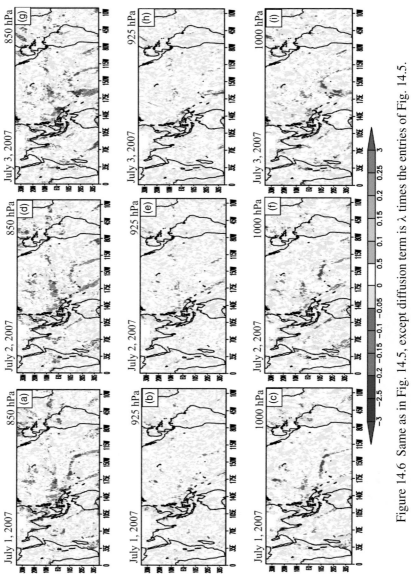

Figure 14.6 Same as in Fig. 14.5, except diffusion term is λ times the entries of Fig. 14.5.

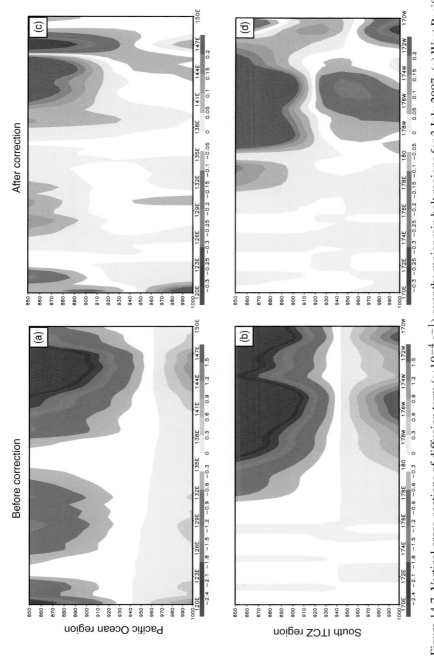

Figure 14.7 Vertical cross-sections of diffusion term ($\times 10^{-4}\ \mathrm{s}^{-1}$) over the major rain belt regions for 3 July 2007: (a) West Pacific Ocean region, averaged over 5°S to 5°N latitude belt; (b) South ITCZ region, averaged over 30°S to 20°S latitude belt; (c) same as (a) but after correction; (d) same as (b) but after correction.

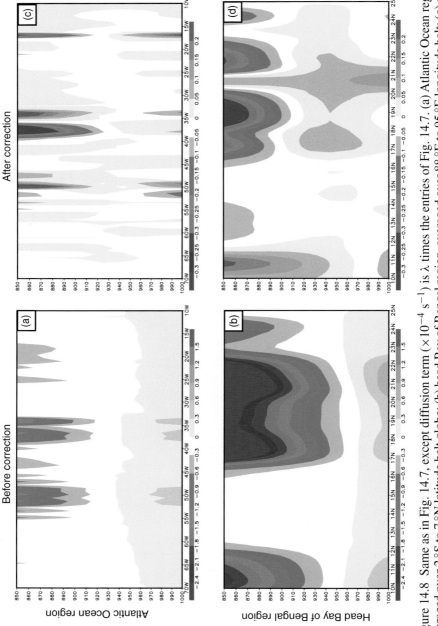

Figure 14.8 Same as in Fig. 14.7, except diffusion term ($\times 10^{-4}$ s^{-1}) is λ times the entries of Fig. 14.7. (a) Atlantic Ocean region, averaged over 2 °S to 7 °N latitude belt globe; (b) head Bay of Bengal region, averaged over 88 °E to 95 °E longitude belt; (c) same as (a) but after correction; (d) same as (b) but after correction.

Table 7.3 *Definition of the PV anomalies used in the study of the 10–12 November 2001 cyclone, associated with mid-upper tropospheric levels (ULev), low tropospheric levels (LLev), and condensational diabatic processes (DIAB)*

PV anomaly	Definition
ULev	PV perturbation above 700 hPa
LLev	Surface thermal anomaly
	and PV perturbation below 700 hPa
DIAB	Positive PV perturbation below 500 hPa in areas with relative humidity exceeding 70%

Strait of Gibraltar at 11 November 00 UTC (Fig. 7.10b). Most of the rainfall and the main wind-induced damage in the Balearic Islands (located in the centre of the western Mediterranean) occurred during this mature phase of the cyclone. The period 11–12 November corresponds to the decay of the disturbance: the low-pressure system filled as it progressed further northwards and the thermal gradient weakened appreciably. Interestingly, the two upper-level embedded disturbances rotated cyclonically about each other as the large-scale parent trough entered the western Mediterranean.

Although the primary role of baroclinic instability is quite clear, the high precipitation potential of this cyclone and the massive cloudy structures that could be observed during the oceanic phase of the storm suggest that condensational latent heat release could have also played a role for the cyclone development. Therefore, diabatically generated PV within the cloud systems will be considered as an explicit factor in the subsequent analysis.

We will consider a simple but physically meaningful decomposition of the perturbation PV field q', which is appropriate to interpret the baroclinic event according to the processes emphasized by the PV thinking theory. The perturbation PV field is defined every 12 h (at 00 and 12 UTC, the available analysis times) during the life cycle of the storm, as the departure from the seven-day time average for the period from 00 UTC on 7 November to 00 UTC on 14 November. We will refer to such reference state as the *MEAN* flow. Three unique anomalies are defined (see Table 7.3): *ULev* is associated with the undulating tropopause and will contain the PV perturbation above 700 hPa plus the θ' field at the upper boundary, located at 100 hPa; *LLev* represents the surface baroclinicity since it is composed of θ' at the bottom boundary (1000 hPa) and the lower-level interior PV perturbation, up to 700 hPa; *DIAB* is expressed as the positive PV perturbation below 500 hPa in areas with greater than 70% relative humidity. The latter is defined to account for the lower-tropospheric PV anomalies associated with condensational heating.[5] Wherever the interior q' is assigned to *DIAB* according to the previous definition, it will be removed from *ULev* or *LLev*. Thus we can refer unambiguously to *ULev* and *LLev* as the dry PV anomalies, associated with the baroclinic processes, and to *DIAB* as a moist anomaly, related to the diabatic contribution of clouds and precipitation. It is important to note that at any point of the interior domain the sum of *ULev*, *LLev* and *DIAB* equal the total perturbation field q', as well as the total θ' field at the boundaries. These anomalies, together with the *MEAN* state, define the total q and θ fields. Note also that *ULev* and *LLev* incorporate both positive and negative PV anomalies (e.g., upper-level troughs and ridges and warm and cold surface anomalies, respectively), whereas *DIAB* is positive everywhere. A further split of the former into their positive and negative pieces (i.e., two additional factors) could be reasonably considered for a more detailed – but obviously more complex – analysis of the event.

The patterns of *ULev*, *LLev* and *DIAB* at the mature phase of the storm are depicted in Fig. 7.11a for some specific isobaric levels. Relevant signatures emphasized previously, such as the two upper-level embedded disturbances, the marked low-level baroclinicity, and the massive cloud formation over the western Mediterranean, are well identified on this PV chart. The inverted balance fields from these anomalies are presented in Fig. 7.11b. Both height deficits (i.e., cyclonic vorticity) and height increases (anticyclonic vorticity) can be found in connection with the dry PV anomalies owing to its double-signed pattern; perturbations of both signs alternate across the whole domain following the synoptic wave train structure. Since the moist *DIAB* anomaly is positive definite it only produces negative height perturbations, essentially in the form of a localized cyclonic vortex over the area of influence of the cyclone. A significant negative height contribution results from each PV anomaly over the area of the surface cyclone (letter C in Fig. 7.11b), thus cooperating with each other to explain the intense barometric depression.

The above PV-inversion results provide a merely static depiction of the cyclogenesis/cyclolysis process, in the sense that only the contributions to the instantaneous cyclone intensity by the selected PV anomalies can be quantitatively diagnosed. Within the framework of PV-thinking theory, it would be illuminating to isolate with a similar degree of detail the true impact of the PV anomalies on the cyclone behaviour (i.e., on changes in its growth, decay, size, trajectory or shape), by means, for example, of the induced surface pressure tendency and vertical motion. More importantly, it would be quite useful to split the contribution of each PV anomaly

---

[5] A subsaturated threshold value of 70% is chosen for two reasons: first, saturated areas are seldom captured in large-scale grid analyses like the ones used in the study, and second, that value allows us to include PV that may advect out of the precipitation region.

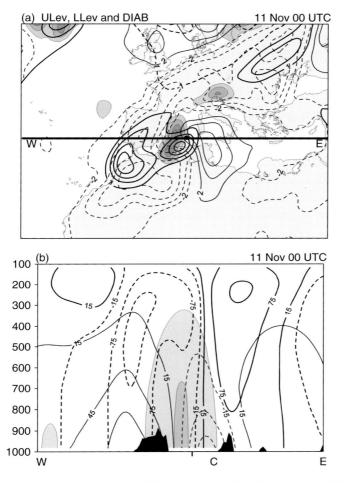

Figure 7.11 (a) Depiction of the PV anomalies defined in the study (*ULev*, *LLev* and *DIAB*; see Table 7.1) on 11 November 00 UTC. *ULev* is shown at 300 hPa using thick contours (solid and dashed lines indicate positive and negative values of the PV anomaly, respectively, starting at 1 and −1 PVU every 1.5 PVU); *LLev* is depicted by means of the thermal boundary condition at 1000 hPa using thin contours (solid and dashed lines indicate positive and negative values of the temperature anomaly, respectively, starting at 2 and −2 °C every 2 °C); *DIAB* is shown at 700 hPa as shaded contours, starting at 0.1 PVU every 0.1 PVU. (b) Vertical cross section along the line WE shown in (a) of the PV-inverted geopotential height perturbation field. Thick contours represent the field inverted from *ULev* (positive and negative values in continuous and dashed lines, respectively, starting at 15 and −15 m every 60 m); thin contours depict the field inverted from *LLev* (positive and negative values continuous and dashed, respectively, starting at 15 and −15 m every 30 m); and shaded contours represent the *DIAB*-inverted field (contours every 30 m starting at −15 m). Letter C at the bottom of the figure indicates the position of the cyclone. (Figure from Romero, 2008)

among its individual effects and the effects produced by its interaction with the other PV features or the mean flow.[6] We could refer to this new perspective as a dynamic approach of the PV-based diagnosis of cyclones. The following method develops such an idea.

7.4.2 PV-based prognostic system combined with the Factor Separation method

A set of prognostic balance equations can be derived from the PV diagnostic system presented in Section 7.3 (Davis and Emanuel, 1991). We start by obtaining tendency equations for (ϕ^t, ψ^t) by taking the local time derivatives of Charney's (1955) nonlinear balance equation (7.2) and the approximate form of Ertel's PV (7.3):

$$\nabla^2 \phi^t = \nabla \cdot f \nabla \psi^t + 2m^2 \left[\frac{\partial^2 \psi^t}{\partial x^2} \frac{\partial^2 \psi}{\partial y^2} + \frac{\partial^2 \psi}{\partial x^2} \frac{\partial^2 \psi^t}{\partial y^2} - 2 \frac{\partial^2 \psi}{\partial x \partial y} \frac{\partial^2 \psi^t}{\partial x \partial y} \right] \quad (7.9)$$

$$q^t = \frac{g \kappa \pi}{p} \left[(f + m^2 \nabla^2 \psi) \frac{\partial^2 \phi^t}{\partial \pi^2} + m^2 \frac{\partial^2 \phi}{\partial \pi^2} \nabla^2 \psi^t \right.$$
$$\left. - m^2 \left(\frac{\partial^2 \psi^t}{\partial x \partial \pi} \frac{\partial^2 \phi}{\partial x \partial \pi} + \frac{\partial^2 \psi}{\partial x \partial \pi} \frac{\partial^2 \phi^t}{\partial x \partial \pi} + \frac{\partial^2 \psi^t}{\partial y \partial \pi} \frac{\partial^2 \phi}{\partial y \partial \pi} + \frac{\partial^2 \psi}{\partial y \partial \pi} \frac{\partial^2 \phi^t}{\partial y \partial \pi} \right) \right]$$
$$(7.10)$$

Recalling that ϕ and ψ are known aspects of the circulation after the inversion of q through Eqs. (7.2) and (7.3), the system (7.9) and (7.10) can be solved for the geopotential and streamfunction tendencies provided q^t is known. This can be calculated using the following form of Ertel's PV tendency equation:

$$q^t = -m(\mathbf{V}_\psi + \mathbf{V}_\chi) \cdot \nabla q - \omega^* \frac{\partial q}{\partial \pi} + \frac{m}{\rho} \boldsymbol{\eta} \cdot \nabla LH \quad (7.11)$$

where the vertical velocity $\omega^* \equiv d\pi/dt$ and irrotational wind $\mathbf{V}_\chi$ must be formally retained (Krishnamurti, 1968; Iversen and Nordeng, 1984), meaning that we cannot simply advect q with the nondivergent wind $\mathbf{V}_\psi$. The horizontal winds are given by the familiar expressions $\mathbf{V}_\psi = m\mathbf{K} \times \nabla \psi$ and $\mathbf{V}_\chi = m\nabla \chi$, and ω^* is related to the more traditional vertical velocity in pressure coordinates (ω) through the expression $\omega^* = (\kappa \pi / p)\omega$. In addition, the only nonconservative effect that is included in this equation is the latent heat release $LH \equiv d\theta/dt$ from nonconvective clouds where

---

[6] The PV field is continuously reshaped by all kinds of lateral and vertical interactions (e.g., Huo *et al.* 1999). From the results of Fig. 7.11 it can be easily inferred, for instance, that the *MEAN* flow will advect all the anomalies essentially eastwards, the *ULev*-induced cyclonic circulation will be responsible for the amplification of the low-level warm anomaly over the western Mediterranean, the *LLev*-induced flow will act advecting the *DIAB* anomaly southwards, etc.

the air is ascending on a broad scale. This diabatic term is parameterized in terms of ω^* as explained in Davis and Emanuel (1991).

The prognostic system is closed by means of an omega equation for ω^* and the continuity equation for the velocity potential χ, which in this context take the following forms:

$$f\eta\frac{\partial}{\partial\pi}\left[\pi^{1-1/\kappa}\frac{\partial}{\partial\pi}(\pi^{1/\kappa-1}\omega^*)\right]+m^2\nabla^2\left(\frac{\partial^2\phi}{\partial\pi^2}\omega^*\right)$$

$$-m^2f\frac{\partial}{\partial\pi}\left(\frac{\partial\omega^*}{\partial x}\frac{\partial\psi}{\partial x\partial\pi}+\frac{\partial\omega^*}{\partial y}\frac{\partial\psi}{\partial y\partial\pi}\right)+\left(f\frac{\partial\eta}{\partial\pi}\frac{1/\kappa-1}{\pi}-f\frac{\partial^2\eta}{\partial\pi^2}\right)\omega^*$$

$$=m^3\nabla^2\left[(\boldsymbol{V}_\psi+\boldsymbol{V}_\chi)\cdot\nabla\theta\right]+mf\frac{\partial}{\partial\pi}\left[(\boldsymbol{V}_\psi+\boldsymbol{V}_\chi)\cdot\nabla\eta\right]-m^2\nabla f\cdot\nabla\left(\frac{\partial\psi^t}{\partial\pi}\right)$$

$$-2m^4\frac{\partial}{\partial\pi}\left[\frac{\partial^2\psi^t}{\partial x^2}\frac{\partial^2\psi}{\partial y^2}+\frac{\partial^2\psi}{\partial x^2}\frac{\partial^2\psi^t}{\partial y^2}-2\frac{\partial^2\psi}{\partial x\partial y}\frac{\partial^2\psi^t}{\partial x\partial y}\right]-m^2\nabla^2 LH \qquad (7.12)$$

$$m^2\nabla^2\chi+\pi^{1-1/\kappa}\frac{\partial}{\partial\pi}(\pi^{1/\kappa-1}\omega^*)=0 \qquad (7.13)$$

The complete system of equations (7.9)–(7.13), formulated in finite differences with vertical staggering of the ω^* levels, is solved iteratively by a simultaneous relaxation method for the fields ϕ^t, ψ^t, q^t, ω^* and χ. It is important to note that the equations are not integrated in time, but simply solved for the instantaneous tendencies. Homogeneous lateral boundary conditions are applied on a large enough domain for each field (i.e., $\phi^t=\psi^t=q^t=\omega^*=\chi=0$), whereas at the top and bottom boundaries more complex conditions are used: the vertical velocity is zero and topographic, respectively; and Neumann type definitions are used for the tendencies ϕ^t and ψ^t:

$$\partial\phi^t/\partial\pi=f\partial\psi^t/\partial\pi=-\theta^t \qquad (7.14)$$

where the necessary potential temperature tendencies at both levels are evaluated from the thermodynamic equation:

$$\theta^t=-m(\boldsymbol{V}_\psi+\boldsymbol{V}_\chi)\cdot\nabla\theta-\omega^*\frac{\partial\theta}{\partial\pi}+LH \qquad (7.15)$$

The prognostic system is solved at 00 and 12 UTC during the life cycle of the Mediterranean cyclone. Of particular interest for describing the cyclogenesis process is the calculated field of surface height tendency around the storm. This field is plotted as unshaded line contours in Fig. 7.12 for 11 November 00 UTC (storm's peak intensity time). As expected, the obtained result identifies the strong cyclogenesis that took place over the western Mediterranean basin at that time

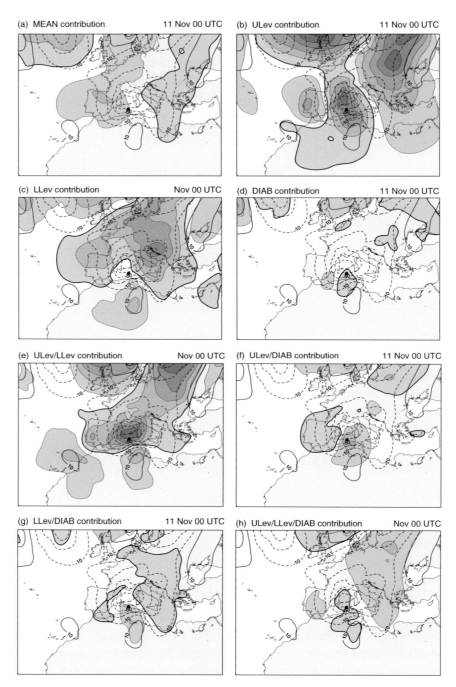

Figure 7.12 Factor separation results on the geopotential height tendency at 925 hPa, corresponding to the mature phase of the cyclone (11 November 00 UTC). The contributions by (a) *MEAN*, (b) *ULev*, (c) *LLev*, (d) *DIAB*, (e) *ULev/LLev*,

Table 7.4 *The eight different atmospheric flow states that are calculated using the PV-based prognostic system for the analysis of the 10–12 November 2001 cyclone. The results are combined in the FS method (see text).*

Flow state	0: *MEAN*	1: *ULev*	2: *LLev*	3: *DIAB*
F_0	Yes	No	No	No
F_1	Yes	Yes	No	No
F_2	Yes	No	Yes	No
F_3	Yes	No	No	Yes
F_{12}	Yes	Yes	Yes	No
F_{13}	Yes	Yes	No	Yes
F_{23}	Yes	No	Yes	Yes
F_{123}	Yes	Yes	Yes	Yes

(recall Fig. 7.10). Since this result emerges from the total balance flow, described by q, ϕ and ψ, we could refer to its signature in Fig. 7.12 as the *total* field for the surface height tendency.

In the present context we are searching for a way to split the total field into the *contributions* from the four constituent PV elements of q (i.e., the basic state *MEAN* plus the three anomalies *ULev*, *LLev* and *DIAB*), including all their possible lateral and vertical interactions. This appears to be a feasible task after the definitions (7.5) and (7.6) based on our ability to repeat the above calculations after subtracting from the input (q,ϕ,ψ) flow state one or several PV anomalies. This is, in essence, a factor separation exercise. With three factors, eight distinct flow configurations are necessary (see Table 7.4). The complete set of effects will be isolated through the algebraic combination of the corresponding solutions to the PV-based prognostic system:

- $E_0 = F_0$
- $E_1 = F_1 - F_0$
- $E_2 = F_2 - F_0$
- $E_3 = F_3 - F_0$

Caption for Figure 7.12 (cont.)
(f) *ULev/DIAB*, (g) *LLev/DIAB* and (h) *ULev/LLev/DIAB* are shown as shaded within thick-line highlighted regions for negative height tendency and shaded for positive height tendency, starting at -5 and $+5$ gpm/12 h, respectively, with a contour interval of 10 gpm/12 h. In all figures, the total height tendency field at 925 hPa is also shown, starting at -10 (dashed line) and $+10$ gpm/12 h (continuous line), with a contour interval of 20 gpm/12 h. The black circle depicted in the southern Mediterranean indicates the central position of the cyclone (see Fig. 7.10). (Figure from Romero, 2008)

- $E_{12} = F_{12} - (F_1 + F_2) + F_0$
- $E_{13} = F_{13} - (F_1 + F_3) + F_0$
- $E_{23} = F_{23} - (F_2 + F_3) + F_0$
- $E_{123} = F_{123} - (F_{12} + F_{13} + F_{23}) + (F_1 + F_2 + F_3) - F_0$

In the following, a more intuitive notation is adopted for these effects by using the terms *MEAN*, *ULev*, *LLev*, *DIAB*, *ULev/LLev*, *ULev/DIAB*, *LLev/DIAB* and *ULev/LLev/DIAB*, respectively.

7.4.3 PV thinking in action: results of the dynamic approach

The impacts and interactions of the potential vorticity anomalies on the surface height tendency are plotted for the mature storm in Fig. 7.12, as spatial patterns with positive and negative signals of the eight different contributions to the total field (this is also included in the figure for reference). Particular attention is paid to signatures of negative height tendency occurring over the western Mediterranean region, thus contributing to the cyclogenesis mechanism.

The action of the background flow *MEAN* is to help progagate the surface cyclone eastwards along the Mediterranean waters (Fig. 7.12a). The individual effect of *ULev* is crucial for the cyclogenesis (Fig. 7.12b). It is responsible for a significant fraction of the surface pressure fall calculated over the western Mediterranean.

Individually, *LLev* represents also a significant contribution to the cyclone growth, movement and decay. During the mature phase its main action is concentrated along the northern flank of the cyclone (Fig. 7.12c), that is, in the zones affected by well-defined surface fronts and where the warm advection is most clear. Thus the pattern of the *LLev* contribution favours the northeastward propagation of the surface disturbance that was emphasized in Section 7.4.1.

The very localized nature of the *DIAB* PV anomaly (Fig. 7.11a) implies a comparatively weak individual effect on the surface height tendency (Fig. 7.12d). This effect is essentially linked to the PV advection induced by the background flow and therefore adopts a dipolar structure about the Algerian coast surrounding the anomaly.

The *ULev/LLev* interaction is a leading agent for generating and driving the November 2001 cyclone (Fig. 7.12e), in agreement with the conceptual model of baroclinic developments formulated by the PV thinking theory. On the surface height tendency this factor attains a comparable or even higher magnitude than the above-mentioned individual contribution of *ULev* (compare with Fig. 7.12b), with the particularity that the pattern over the western Mediterranean very closely resembles the total tendency field. The importance of the synergistic action is well understood bearing in mind the appreciable warm advection values that are induced at low levels.

The *ULev/DIAB* effect is quite relevant (Fig. 7.12f). It displays a dipolar structure like the one noted for *DIAB* (Fig. 7.12d), except of larger magnitude and reversed sign. The sign reversal is due to the opposite directions of the *MEAN* and *ULev* advective winds acting, respectively, on the *DIAB* PV anomaly (not shown). In all cases, the *ULev/DIAB* factor is contributing to cyclogenesis only to the west of the cyclone; over the cyclone position and eastwards, its effect is essentially cyclolytic (Fig. 7.12f).

The *LLev/DIAB* synergism is also cyclolytic over the mature cyclone (Fig. 7.12g). Since the *LLev/DIAB* mechanism incorporates the effects of the *DIAB* PV advection by the *LLev*-induced flow and the *LLev* thermal advection by the *DIAB*-induced flow, both of comparable magnitude, the precise interpretation of the *LLev/DIAB* spatial pattern displayed on Fig. 7.12g appears to be more intricate than for the other factors.

Finally, we can analyse the contribution of the triple interaction *ULev/LLev/DIAB* (Fig. 7.12h). Owing to the involvement of the small *DIAB* PV anomaly in this factor, it exerts only a secondary role, like *DIAB*, *ULev/DIAB* and *LLev/DIAB*. Interestingly, the obtained height tendency pattern almost perfectly balances with the *LLev/DIAB* contribution (compare with Fig. 7.12g), helping in this case to deepen the cyclone.

In summary,[7] it can be concluded that the most intense phase of the November 2001 Mediterranean cyclone was regulated by the *MEAN*, *ULev*, *LLev* and *ULev/LLev* processes: *MEAN* and *LLev* assisted in the northeastward propagation of the disturbance and to some extent to its intensification, especially *LLev* during its African phase; both *ULev* and *ULev/LLev* were fundamental for the African cyclogenesis and further intensification of the disturbance over the western Mediterranean, although *ULev/LLev* quickly decayed after the cyclone maturation. The remaining mechanisms (*DIAB*, *ULev/DIAB*, *LLev/DIAB* and *ULev/LLev/DIAB*) were most relevant during the mature phase of the system but, compared to the first group, they only exerted a secondary role owing to the limited strength and spatial dimensions of the *DIAB* PV anomaly. According to the factor separation results, the development and mature phases of the baroclinic disturbance can be described as a meteorological scenario of effective cooperation among the background flow and the PV anomalies over the surface cyclone domain, whereas the decay of the system is linked to a scenario of much weaker or even cyclolytic interactions between some of these factors.

---

[7] For a deeper discussion, which also includes results for the development and decay phases of the baroclinic sytem, see Romero (2008).

8

Experience in applying the Alpert–Stein Factor Separation Methodology to assessing urban land-use and aerosol impacts on precipitation

S. C. van den Heever, C. Rozoff, and W. R. Cotton

In this chapter, Alpert–Stein Factor Separation (FS) Methodology is used to assess the significance of various land surface characteristics on the development and precipitation characteristics of convective storms occurring downwind of an urban region. In particular, the roles of topography, momentum fluxes, radiative heat fluxes, and latent and sensible heat fluxes are evaluated. The use of this technique in investigating the relative and interactive roles of different nucleating aerosols, including cloud condensation nuclei, giant cloud condensation nuclei, and ice nuclei, on the development, structure, and precipitation processes of tropical convection is also then described.

8.1 Numerical mesoscale simulations of urban enhanced convection

8.1.1 Introduction

We briefly review the application of a three-dimensional, cloud-resolving, mesoscale model case study described in Rozoff *et al.* (2003) to the examination of the impacts of St. Louis, MO, USA, land use and topography on local convective storms. Located within a relatively moist and temperate climate, St. Louis is an ideal city for experimental study since it is relatively isolated from other substantial urban areas and its local geography is devoid of major topography and large bodies of water. Furthermore, St. Louis was the site for a large field campaign in 1971–5, called Project METROMEX (Changnon *et al.*, 1981). That study was dedicated to questions regarding the role of urban areas on weather modification. It is believed the findings for St. Louis are widely applicable to other cities containing similar background conditions. Using Alpert–Stein Methodology, Rozoff *et al.* examined the impacts of increased momentum drag of the urban surface, the urban heat island, and local topography on convective storms around St. Louis.

Factor Separation in the Atmosphere: Applications and Future Prospects, ed. Pinhas Alpert and Tatiana Sholokhman. Published by Cambridge University Press. © Cambridge University Press 2011.

8.1.2 Model description

The Regional Atmospheric Modeling System (RAMS) (Cotton *et al.*, 2003), a nonhydrostatic cloud-resolving mesoscale model, was used with three interactive, nested grids in order to both adequately capture the large-scale environment and focus in on convective-scale motions in the immediate St. Louis region. Figure 8.1 shows the three grid locations. The large, intermediate, and inner grids contained 37.5, 7.5, and 1.5 km grid spacing, respectively. A stretched vertical coordinate was incorporated so that lower levels, particularly within the boundary layer, had higher vertical resolution. Specifically, 8 of the 40 vertical levels roughly encompassing 22 km of atmosphere were contained in the lowest 1 km.

The RAMS surface parameterization, known as the Land Ecosystem–Atmosphere Feedback model (LEAF-2, Walko *et al.*, 2000) was used in this study. LEAF-2 contains 30 distinct land/water classes each containing distinct properties including albedo, emissivity, leaf area index, vegetation fractional coverage, roughness length, and displacement height. These parameters were utilized in the calculations of the surface radiation, turbulent sensible/latent heat and momentum fluxes, which ultimately communicate properties of land use and land cover to the atmosphere. However, to account for the three-dimensional geometry and the efficient heat storage of urban surfaces, a more sophisticated surface parameterization

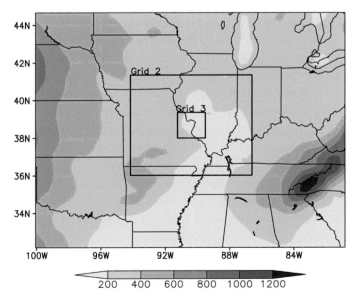

Figure 8.1 The three boxes pictured above represent the boundaries of each grid described in the text. The background field is the topography (m) (after van den Heever and Cotton, 2007).

known as the Town Energy Budget (TEB) (Masson, 2000) was used in regions of urban land use. Each grid cell allows for three distinct land-use and land-cover types, which were distributed proportionately according to the land-use and land-cover input data.

8.1.3 Case study results

The selected case study examines widespread, local convective storms that occurred in St. Louis on 9 June 1999. On this date, the large-scale weather pattern surrounding St. Louis lacked obvious forcing mechanisms for convective initiation. However, the warm and moist environment was favorable for convection. The relatively gentle vertical shear of the horizontal wind favored outflow-dominant storms. The lack of obvious large-scale forcing mechanisms made this case study a promising candidate for the investigation of the potential role of surface impacts on convective initiation and evolution.

Before describing the simulated evolution of convection, it is worth mentioning the simulated mesoscale flows in the local vicinity of St. Louis prior to storm initiation. A notable heat island of nearly 2 °C formed by 1800 UTC (1100 local time). In addition, water vapor mixing ratios were consistently 1 to 2 g kg^{-1} lower over St. Louis during the late morning hours. Confluence of the surface wind over and downwind of St. Louis increased throughout the morning.

By 1900 UTC, deep, moist convective clouds formed in the simulation. Figure 8.2 shows the vertically integrated, liquid-equivalent condensate mixing ratio (which accounts for cloud water, rain, pristine ice, snow, aggregates, graupel, and hail) and total precipitation for the first 6 h of convective storms in the simulation. Storms were fairly widespread throughout the simulation, but the locations of initial storm formation are of primary interest. Around 1900 UTC, a storm developed southwest of St. Louis while two storms appeared to be forming on the periphery of the dense urban land use in St. Louis. A full storm appeared downstream of St. Louis over the area of concentrated low-level confluence by 2000 UTC and resulted in approximately 80 mm of accumulated precipitation downwind of St. Louis. Incidentally, radar data indicated that a storm formed in nearly the same area at a slightly earlier time, although it is not immediately obvious that available observations can confirm whether the urban surface played a role in the observed storms. Later in the simulation (Fig. 8.2c–f), the storms merged into a concentrated cluster of individual cells that moved southward against the mean flow.

8.1.4 Sensitivity experiments

An ensemble of sensitivity experiments were performed in order to apply the Alpert–Stein Factor Separation Methodology to the urban convection problem. Simulations

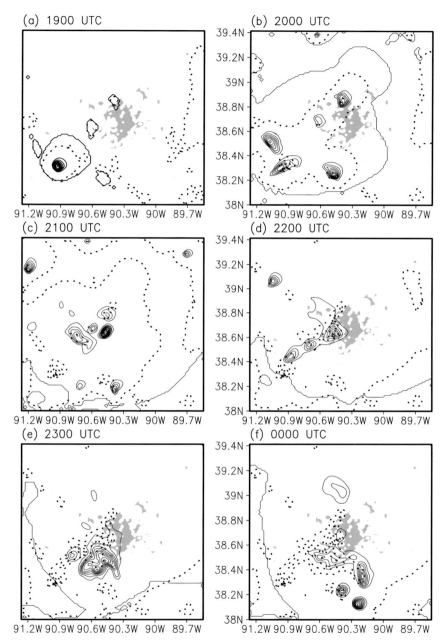

Figure 8.2 Hourly vertically integrated condensate (solid contours, with 5 mm intervals) and total precipitation (dashed contours, with 40 mm intervals). Urban land use is shaded in the background (after Rozoff *et al.*, 2003).

were performed with and without urban surface radiative and latent/sensible heat fluxes, urban surface momentum flux, and local topography, yielding eight total experiments.

In simulations where urban parameters and the TEB scheme were ignored in calculations of either the surface radiation, latent heat, and sensible heat fluxes or the momentum flux, the parameters from surrounding rural land use and the LEAF-2 surface scheme were implemented instead. In particular, the LEAF-2 parameters for cropland and wooded grassland were used in flux calculations ignoring the urban and suburban land uses, respectively. While rural heterogeneity likely influences convection as well, this experimental design helps isolate the role of urban land use on the atmosphere. Finally, special care must be taken when isolating the role of topography on convective storms. An ideal sensitivity experiment should neglect localized impacts of topography while avoiding detrimental impacts on the synoptic scale circulation. Therefore, fine-scale topographic details in the inner grid can be neglected by interpolating the outer grid topography into the inner grids. The topographic details are shown in Fig. 8.3. The eight experiments are summarized in Table 8.1. Experiment E_{123} is the simulation described in Section 8.2.3 and is also referred to as the control experiment.

Before discussing the role of local topography and St. Louis on the evolution of convective storms, it is important to understand the results of experiment E_0 with the aim of grasping background factors in play. Without local topography and urban effects, storm initiation arises from other factors such as variability in the

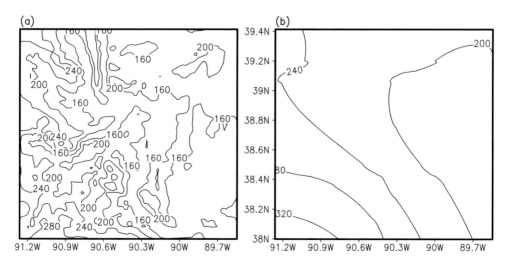

Figure 8.3 (a) Regular and (b) smoothed topography in the inner grid over St. Louis (after Rozoff et al., 2003).

Table 8.1 *Summary of sensitivity experiments performed. The "Rural" and "Urban" specifications indicate which type of surface layer calculation was used.*

Experiment	Radiation/heat fluxes	Topography	Momentum flux
E_0	Rural	Smoothed	Rural
E_1	Urban	Smoothed	Rural
E_2	Rural	Unsmoothed	Rural
E_3	Rural	Smoothed	Urban
E_{12}	Urban	Unsmoothed	Rural
E_{23}	Rural	Unsmoothed	Urban
E_{13}	Urban	Smoothed	Urban
E_{123}	Urban	Unsmoothed	Urban

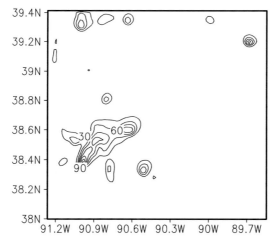

Figure 8.4 Total precipitation (mm) for experiment E_0 at 0000 UTC 9 June 1999. The contour interval is 30 mm (after Rozoff *et al.*, 2003).

vegetated land-use types. The results of the experiments indicate storms initiate an hour later in E_0 than in E_{123}. This is perhaps not surprising since experiment E_0 does not possess as much of the control experiment's surface variability. Figure 8.4 provides the total precipitation through the 24 h simulation period ending 0000 UTC 9 June. Heavy precipitation totals are found in localized clusters, indicating the isolated nature of the simulated storms and the relatively small amount of storm movement. Unlike in the control experiment, E_{123}, no precipitation falls downwind of St. Louis in E_0.

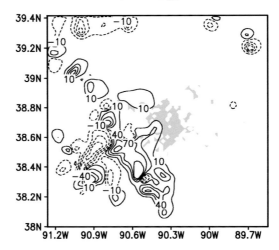

Figure 8.5 The difference in total precipitation (mm) between experiments E_2 and E_0 at 0000 UTC 9 June 1999. The contour interval is 30 mm. Negative contours are given by dotted lines (after Rozoff *et al.*, 2003).

Contribution by topography

We first explore how local topography around St. Louis changes the results in E_0. This analysis is facilitated through simple difference fields between experiments E_2 and E_0. Figure 8.5 illustrates the difference in accumulated precipitation at 0000 UTC 9 June 1999. It is clear that the significant precipitation southwest of St. Louis in E_0 is displaced toward the northeast and southwest. Furthermore, precipitation formed an hour earlier (initially to the southwest of St. Louis) and the area-average total precipitation is 2.4 mm greater in E_2.

The difference field for total precipitation shows the net impact of topographic details on the outcome of convective storms. Adding enhanced topographic details to the inner grids creates a much more heterogeneous field of convective available potential energy (CAPE) and low-level flow, especially in the region of the greatest topographic variability (not shown) south and west of St. Louis. Thus, it appears more complex topography and the resulting flow response offer more concentrated pockets of forcing and instability favorable to convective initiation. The enhanced convective activity also increases the number of cold pools that force subsequent convection. The greater opportunities for convective initiation appear to explain the larger net precipitation in E_2.

Contribution by radiative and latent/sensible heat fluxes

Experiment E_1 adopts rural formulations for both the calculation of surface momentum flux and the smoothed topography as in E_0, but the suburban and urban land-use

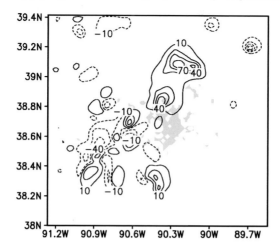

Figure 8.6 The difference in total precipitation (mm) between experiments E_1 and E_0 at 0000 UTC 9 June 1999. The contour interval is 30 mm. Negative contours are given by dotted lines (after Rozoff *et al.*, 2003).

regions revert to the urban formulation for the surface layer radiation and the turbulent sensible and latent heat fluxes. The differences between E_1 and E_0 essentially demonstrate contributions to convective activity from the urban heat island. However, as will be shown later, the strength of the urban heat island is indirectly related to the surface momentum flux as well.

The difference field for the total accumulated precipitation at 0000 UTC 9 June (Fig. 8.6) immediately shows a possible urban signal downwind of St. Louis. Tracing the actual evolution of convection in E_1 confirms convective initiation just downwind of the city at 1900 UTC. However, this downwind convection actually forms about an hour earlier than the downwind convective initiation in the control simulation. Elsewhere, convection appears to be randomly rearranged from E_0 to E_1. As before, the increased area-average total precipitation in experiment E_1 (1.5 mm more) looks as if the increased surface forcing generates more convection than in E_0.

While the precipitation pattern in experiment E_1 is suggestive of urban impacts on convective initiation, more compelling physical insight is achieved by looking at the difference fields of other variables. Figure 8.7a provides low-level difference fields for temperature, water vapor mixing ratio, and wind vectors at 1800 UTC. A substantial heat island of nearly 2 °C is clearly evident in E_1 at this time, which is slightly displaced downstream (north) of St. Louis. At the same time, water vapor mixing ratios are depressed in this heat island. These changes result directly from an enhanced (weakened) sensible (latent) heat flux. The heat island also directs

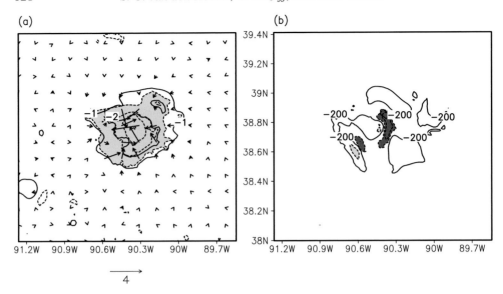

Figure 8.7 (a) Difference fields, between E_1 and E_0, of low-level (48 m) temperature (dashed contour at every 0.5 °C; shading indicates regions with temperature
differences exceeding 0.5 °C), water vapor mixing ratio (solid contour every
$1\,\mathrm{g\,kg^{-1}}$), and winds ($\mathrm{m\,s^{-1}}$ with reference vector at bottom of panel) at 1800
UTC. (b) Difference fields showing 48 m divergence (dashed contour at every
$0.5 \times 10^{-3}\,\mathrm{s^{-1}}$; dark and light shading indicate divergence below -0.5 and above
$+0.5 \times 10^{-3}\,\mathrm{s^{-1}}$, respectively) and surface-derived CAPE (solid contour; $\mathrm{J\,kg^{-1}}$)
(after Rozoff *et al.*, 2003).

significant low-level flow into the heat island. Difference fields of low-level divergence and surface-based CAPE are provided in Fig. 8.7b. As expected from the
low-level wind field, strong convergence appears near the center of the urban heat
island. Some positive divergence indicating subsidence is found on the outskirts
of the urban area, particularly to the southwest of the urban heat island. The flow
patterns appear to directly influence the horizontal distribution of CAPE as well.
Because of decreased low-level moisture, pockets of lower CAPE dominate much
of the heat island region. Still, this region is convectively unstable in E_1 with CAPE
values around 2400–2600 $\mathrm{J\,kg^{-1}}$.

Contribution by urban momentum fluxes

Experiment E_3 isolates the role of the urban surface momentum flux in convective
storms. Here we test the hypothesis that the urban surface creates enhanced drag on
the low-level environmental flow, resulting in sufficient convergence for convective
initiation. As it turns out, convection is not significantly impacted in the immediate
vicinity of St. Louis in this experiment. While increased drag slows the low-level

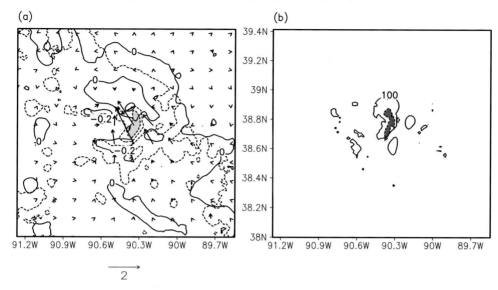

Figure 8.8 (a) Difference fields, between E_{12} and E_{123}, of low-level (48 m) temperature (dashed contour at every $0.2\,°C$; dark and light shading indicate regions with temperatures below $-0.2\,°C$ and above $+0.2\,°C$, respectively), water vapor mixing ratio (solid contour every $1\,g\,kg^{-1}$, and winds ($m\,s^{-1}$ with reference vector at bottom of panel) at 1800 UTC. (b) Difference fields showing 48 m divergence (dashed contour at every $0.5 \times 10^{-3}\,s^{-1}$; dark and light shading indicate divergence below -0.5 and above $+0.5 \times 10^{-3}\,s^{-1}$, respectively) and surface-derived CAPE (solid contour; $J\,kg^{-1}$) (after Rozoff *et al.*, 2003).

wind by around $1\,m\,s^{-1}$, the convergence (divergence) on the windward (leeward) sides of St. Louis is an order of magnitude smaller than the values found in E_1 or E_2. The resulting convective environment evolves in a way that resembles E_0.

To explore feedbacks between the surface radiative/heat fluxes and the surface drag, the differences between an experiment replacing the urban formulation of surface momentum flux with a rural formulation (E_{12}) and the control experiment (E_{123}) are now considered. Figure 8.8a provides the 1800 UTC difference fields between E_{12} and E_{123} of temperature, water vapor mixing ratio, and winds at the lowest model level. The subtraction of enhanced urban drag leads to an invigorated urban heat island downwind of St. Louis. However, temperatures slightly decrease on the windward side of the city. The wind response is notable, with enhanced flow over the city. A significant number of the changes in low-level wind speeds are merely a consequence of decreased surface drag on the flow. However, the enhanced urban heat island also contributes to the increased flow in E_{12}. Surface divergence and CAPE differences are provided in Fig. 8.8b, which indicates that an enhanced urban heat island and reduced drag strengthens low-level convergence on the leeward edge of St. Louis. This area contains slightly greater CAPE, which is primarily a result of enhanced low-level temperatures. Similar to experiment E_1, a storm forms downwind of St. Louis 1 h earlier in E_{12} than in the control experiment.

An understanding of why the surface turbulent momentum flux impacts the strength of the urban heat island and the timing of convective initiation is found in the details of the surface sensible heat flux formulation. If we consider a grid cell within St. Louis that is completely composed of urban land use, the sensible heat flux $H_\alpha(\mathrm{W\,m^{-2}})$ in the TEB scheme is

$$H_\alpha = \rho c_p C_\alpha (T_{s,\alpha} - T_a),$$

where α denotes the type of surface (i.e., roof, wall, road), ρ is the air density $(\mathrm{kg\,m^{-3}})$, c_p is the specific heat of dry air at constant pressure $(\mathrm{J\,kg^{-1}\,K^{-1}})$, C_α is the drag coefficient $(\mathrm{s^{-1}})$, $T_{s,\alpha}$ is the temperature (K) at the surface of α, and T_a is the air temperature (K) for the first model level (or canopy temperature if α represents a wall or road surface within the urban canyon formulation of TEB). It is important to note that the drag coefficient C_α is a nonlinear function of wind speed, temperature, and the water vapor mixing ratio.

Using the sensible heat flux formulation of TEB and the meteorological output from the simulations within St. Louis, the relative contribution to increased sensible heat flux in experiment E_{12} by the wind speed and by the relevant thermodynamic variables (temperature and humidity) may be determined. Figure 8.9a provides the sensible heat flux over a 5 h period prior to convective initiation for experiments E_{12} and E_{123}. The rural formulation for momentum flux results in a greater sensible heat flux over St. Louis. According to the TEB scheme, the enhanced sensible heat flux creates cooler surface temperatures in E_{12}, although temperatures at 48 m

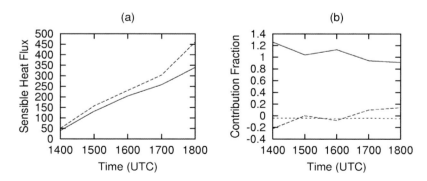

Figure 8.9 (a) Sensible heat flux $(\mathrm{W\,m^{-2}})$ for experiments E_{12} (dashed) and E_{123}(solid). (b) The fractional contribution of wind speed changes (solid), thermodynamic changes (long dashed), and the residual term (short dashed) to total sensible heat flux differences (after Rozoff *et al.*, 2003).

(the first model level) slightly increase, as was seen in Fig. 8.8a. We can split the total difference in sensible heat flux between E_{12} and E_{123} into a sum of several differences, including a difference in sensible heat resulting from a difference in wind speed when temperature and humidity are fixed, a difference in temperature and humidity while wind speed is fixed, and a residual term. Dividing each difference by the total difference in sensible heat provides a fractional measure of how much wind speed and both the temperature and humidity each contribute to the total difference in urban sensible heat flux between E_{12} and E_{123}. Figure 8.9b provides the fractional contributions of each factor, showing that the differences in sensible heat flux largely result from the greater wind speed in E_{12}. As the surface layer becomes increasingly unstable in E_{12}, the enhanced temperature starts to play a positive role in enhancing the sensible heat flux in E_{12} over E_{123}. Finally, it should be mentioned that the residual term is decidedly insignificant, accounting for only -4% of the difference in sensible heat flux.

Interactions between factors

As just seen, the sensitivity experiments E_0, E_1, E_2, and E_3 only provide a partial explanation for what is seen in the control experiment E_{123}. In order to get a complete picture of the convective evolution in E_{123}, we must consider the interactions between the three factors. Rozoff *et al.* (2003) carefully considered all interaction fields in the eight sensitivity experiments. In summary, they found that interactions between surface urban radiation/heat fluxes and topography and interactions among all three factors led to a slight decrease and increase, respectively, in the initial development of convection downwind of St. Louis. The interactions between urban momentum flux and topography did not have a significant impact on convective initiation. Finally the impact of interactions between urban radiation/heat fluxes and momentum flux showed the most significant change in downwind convective initiation. The interactions between urban surface fluxes apparently account for the delayed downwind convection in E_{123}, which occurs almost an hour later than in E_1. The cause for this discrepancy is answered in the previous section discussing the feedbacks that occur between the urban surface radiative/heat fluxes and urban momentum flux. Increasing the surface drag decreases low-level winds. The weaker winds consequently reduce the sensible heat flux and the strength of urban heat island convergence downwind of St. Louis

8.1.5 Summary and discussion

The Alpert–Stein Factor Separation Methodology was applied to an idealized set of sensitivity experiments to examine the impact of local topography, surface radiative and heat fluxes, surface turbulent momentum flux, and their interactions on convective initiation and evolution. It was concluded that topographic details do play a significant role in initiating convection, particularly south and west of St. Louis, where topographic variability is more substantial. The radiative and turbulent latent/sensible heat fluxes from the urban surface result in a significant daytime urban heat island that facilitates downwind convective initiation in an area of enhanced low-level ascent. While these particular numerical experiments show that urban surface drag does not produce sufficient low-level convergence for convective initiation in this case study, interactions between momentum flux and the sensible heat flux play a significant role in the strength of the urban heat island and the timing of convective initiation. Namely, increased surface drag over St. Louis slows the low-level winds, which decreases the transfer of surface heat into the boundary layer. The heat island circulation is therefore slightly weakened in the presence of stronger drag, delaying the development of a downwind convective storm.

The analysis of interactions via the Alpert–Stein Methodology is useful in uncovering important relationships between key variables in the urban convection problem. It is important to note, however, that while the factor analysis can indicate the outcome of interactions, the physical processes that lead to the difference fields between sensitivity experiments still need to be examined closely. The use of the Alpert–Stein Methodology in future numerical studies of the urban impacts on convective storms, whether such work investigates a particular case study for a city or an ensemble of storm situations for a particular city, is highly recommended. Also, future studies may benefit from a more general approach to factor separation by Krichak and Alpert (2002), which includes both a traditional factor separation analysis of numerical simulations and a way to understand the interactions between the background state and the physical factors being studied. For instance, it may be interesting to know how the factor interactions and simulation differences change by varying surface parameters, such as urban roughness length, over a wider range of values.

8.2 Impacts of aerosols on the precipitation characteristics of convective storms over Florida

8.2.1 Introduction

Aerosols, whether natural or anthropogenic, affect the climate both directly through the scattering and absorption of radiation, and indirectly by modifying the cloud properties that then, in turn, influence the radiation budget through what is known as cloud radiative forcing. Two aerosol indirect effects are typically considered. The *first indirect effect* (Twomey, 1974) argues that, for the same water content, increasing the concentrations of cloud condensation nuclei (CCN) leads to a greater number of activated cloud droplets, higher cloud droplet concentrations, an increase in the cloud optical depth, an increase in the albedo, and hence more-reflective clouds. For low-level clouds, which radiate at similar temperatures to the Earth, this will tend to cool the planet. Estimates suggest that this effect may play a significant role in counteracting the "greenhouse effect." *The second indirect effect* (Albrecht, 1989) postulates that when the number of CCN is increased, cloud droplet sizes are decreased, the cloud droplet size spectra are narrowed, and the formation of rain droplets through collision and coalescence of cloud droplets is thus reduced, thereby suppressing precipitation processes. This leads to longer cloud lifetimes, enhanced cloud fractions, and hence more reflective clouds as the cloud droplets are smaller and the liquid water path is greater. While the first and second indirect effects are relatively easy to understand conceptually, recent research has indicated that cloud and precipitation responses may not always be in keeping with these hypotheses, and that they may vary with cloud type, cloud dynamics, storm life cycle, and the type of nucleating aerosol being considered.

Urban and industrial pollution, smoke from forest fires and other burning vegetation, and dust are all sources of CCN, giant CCN (GCCN) and/or ice nuclei (IN). Should such sources of aerosol be ingested into storms, the greater concentrations of CCN associated with urban pollution and smoke particles act to inhibit precipitation processes, and may even completely shut off precipitation at times through aerosol indirect forcing. This has been seen in both observational and modeling studies (Kaufman and Nakajima, 1993; Borys *et al.*, 1998; Rosenfeld and Lensky, 1998; Rosenfeld, 1999, 2000; Andreae *et al.*, 2004; van den Heever *et al.*, 2006; van den Heever and Cotton, 2007; Berg *et al.*, 2008).

The effects of GCCN on cloud characteristics and precipitation processes have received relatively less attention compared with CCN. Unlike CCN, GCCN appear to enhance precipitation (Hobbs *et al.*, 1970; Eagan *et al.*, 1974; Dettwiller and Changnon, 1976; Hindman, 1977a,b; Braham *et al.*, 1981; Mather, 1991) by accelerating the formation of droplets large enough to initiate the collision and coalescence process. Feingold *et al.*'s (1999) simulations showed that the presence of GCCN can change nonprecipitating stratocumulus clouds into precipitating clouds, and that the relative impact of GCCN on the formation of drizzle increases with increasing CCN concentrations. Laird *et al.* (2000) demonstrated that the efficiency of the initial development of precipitation in small Florida cumuli was enhanced by the presence of ultragiant nuclei. A modeling study (van den Heever and Cotton, 2007) of convective storms developing downwind of St Louis, MO, demonstrated enhanced surface precipitation associated with increased GCCN concentrations produced by the city, and recent observations from a satellite study conducted on a global scale suggest that maritime clouds that form in regions of enhanced sea salt concentrations (GCCN) tend to form larger raindrops and precipitate more frequently (L'Ecuyer *et al.*, 2009).

Dust, a natural aerosol, is transported from major desert regions all around the globe, thereby influencing the aerosol composition and concentrations in locations far removed from their source (Prospero 1996, 1999). It has been known for some time from laboratory studies and measurements that dust can serve as efficient IN (Schaefer, 1949; 1954; Isono *et al.*, 1959; Gagin, 1965; Roberts and Hallett, 1968; Levi and Rosenfeld, (1996); Zuberi *et al.*, 2002; Hung *et al.*, 2003). More recently, Sassen *et al.* (2003) showed using aircraft and lidar data that a mildly supercooled altocumulus cloud over southern Florida was glaciated in the presence of Saharan dust, which acted as effective IN. Extremely high concentrations of IN have also been found within dust layers over Florida, confirming the efficiency of dust aerosols in the nucleation of ice (DeMott *et al.*, 2003). Levin *et al.* (1996) demonstrated through modeling studies that dust can become coated with sulfates and hence serve not only as IN, but also as CCN and GCCN. This is supported by the satellite observations of Rosenfeld *et al.* (2002) that show clouds in polluted air are nonprecipitating over land but rapidly transform into precipitating clouds once the airmass advects over the sea and sea-salt particles (GCCN) become entrained into the clouds. Finally, it has recently been demonstrated that since small amounts of soluble or slightly soluble material are often naturally present in soil-derived particles, dust does not necessarily have to undergo atmospheric processing in order to be an efficient condensation nucleus and this, together with the relatively large size of dust particles, means that even fresh dust can be important in warm cloud formation (Twohy *et al.*, 2009).

Dust and other aerosols, whether natural or anthropogenic, can therefore substantially alter the characteristics of clouds and cloud processes in a variety of ways, and hence influence the accumulated precipitation at the surface. However, there is still no general consensus, whether quantitative or qualitative, on the impacts of CCN, GCCN, and IN on cloud processes and precipitation rates. This is further complicated by the fact that the influences of one of these nucleating species may offset the impacts of another species. For example, when the concentrations of both CCN and GCCN are enhanced within a cloud, will precipitation be enhanced due to the presence of GCCN or be suppressed due to the presence of enhanced CCN concentrations? Which species is going to exert the dominant influence? Another aspect to consider is whether the impacts of these species exist independently of one another, or do they interact, and if so, is this interaction constructive or destructive? These questions are further complicated when the impacts of IN also need to be taken into account in mixed phase clouds.

The results presented in this chapter focus primarily on the use of the Alpert–Stein Factor Separation Methodology as used by van den Heever *et al.* (2006). Further details of the study can be found in that paper. The simulations are based on observations acquired during the CRYSTAL-FACE field program (Jensen *et al.*, 2004). During the field campaign, high aerosol concentrations were observed over Florida on 28 and 29 July which were associated, certainly in part, with the penetration of Saharan dust over the peninsula during this time (DeMott *et al.*, 2003; Sassen *et al.*, 2003). Our primary goal is to investigate the impacts that variations in the concentrations of CCN, GCCN, and IN, associated with Saharan dust intrusions, have on the precipitation produced by storms over Florida.

8.2.2 Model and experiment setup

Model configuration

The RAMS cloud-resolving model (Pielke *et al.*, 1992; Cotton *et al.*, 2003) was also used for this study. Four nested model grids with horizontal grid spacings of 50, 10, 2, and 0.5 kilometers were employed, the locations of which are shown in Fig. 8.10. Thirty-six vertical levels with variable grid spacing were used, and the model top extended to approximately 20 km above ground level (AGL). The model fields were initialized using 40 km ETA data. These data also served as boundary conditions throughout the simulation. Convection was explicitly resolved on grids 3 and 4. For a detailed description of the model setup and parameterization schemes

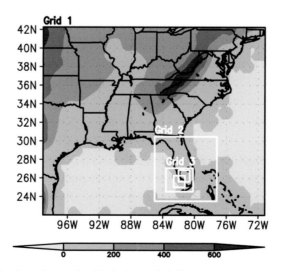

Figure 8.10 The locations of grids 1 through 4. Due to computer constraints, grid 4 was limited in size and needed to be moved during the simulation. Both locations of grid 4 are indicated (after van den Heever *et al.*, 2006).

used the reader is referred to van den Heever *et al.* (2006), although some of the microphysical details particularly relevant to this discussion are described here.

The mass and number concentrations of the various hydrometeors were predicted through the use of a two-moment bulk microphysical scheme (Meyers *et al.*, 1997), and all seven of the available hydrometeor species (cloud droplets, rain, pristine ice, snow, aggregates, graupel, and hail) were activated. The size distribution spectra of these hydrometeor species are represented using a generalized gamma distribution function. The cloud droplet size spectrum is decomposed into two modes, the first representing cloud droplets with diameters between 1 and 40 micrometers, and the second for diameters between 40 and 80 micrometers. This bimodal cloud droplet spectrum is more representative of observed cloud droplet distributions and allows for a more accurate representation of observed cloud droplet collection. In RAMS, the number of cloud droplets or pristine ice crystals ($N_{\text{activated}}$) formed is given by the general formula:

$$N_{\text{activated}} = N_{\text{available}} F_{\text{activation}}$$

where $N_{\text{available}}$ is the maximum number of CCN, GCCN, or IN that are available for activation of the first and second cloud modes and pristine ice respectively, and $F_{\text{activation}}$ is an activation function based on ambient conditions. For the first cloud mode, $F_{\text{activation}}$ is a function of vertical velocity, temperature, water supersaturation, the size and chemistry of the aerosol, while for the second cloud mode, $F_{\text{activation}}$ is simply dependent on water supersaturation. The number of pristine ice crystals formed by deposition and condensation-freezing of IN is determined using a generalization of the Meyers *et al.* (1992) formula and is a function of ice supersaturation. The variable $N_{\text{available}}$, representing the concentration of either CCN, GCCN, or IN, is the model's aerosol prognostic variable for each of these species, and is initialized either heterogeneously or homogeneously, has sources and sinks, and is advected and diffused.

Experiment design

On 28 July 2002, a high dust day during the CRYSTAL-FACE field campaign, an easterly wave penetrated over the southern Florida peninsula and transported Saharan desert dust to this region. Storms initially developed to the south and southwest of Lake Okeechobee, enhanced by the presence of the easterly wave (Fig. 8.11). These storms continued to strengthen, accompanied by the development of new convection, and progressed in a westward to southwestward direction, reaching the coastal regions of Everglade City and Fort Meyers between 2100 and 2200 UTC. Convection started to weaken after ~2200 UTC, leaving the remnant anvils extending over the oceans for several more hours.

(a) 19:02 UTC

(b) 20:02 UTC

(c) 21:15 UTC

(d) 22:02 UTC

(e) 23:02 UTC

(f) 23:45 UTC

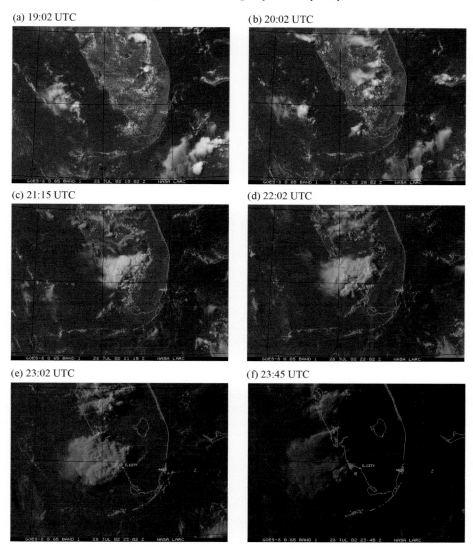

Figure 8.11 Visible satellite imagery of the storm development over the Florida peninsula at approximately hourly intervals on 28 July 2002 (after van den Heever *et al.*, 2006); figures used with the permission of Louis Nguyen, NASA Langley Research Center).

In order to simulate the impacts of varying aerosol concentrations on the convective development and resultant anvils, idealized vertical profiles of the concentrations of the aerosols available for activation were established based on measurements made on relatively "clean" and "dusty" days during the field campaign. EXP-CLN and EXP-OBS will be used to designate these cases, respectively.

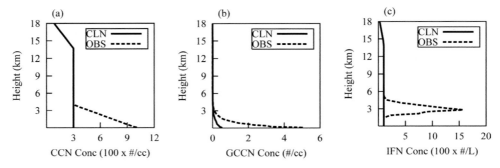

Figure 8.12 Idealized vertical profiles of the available aerosol that can serve as (a) CCN, (b) GCCN, and (c) IN profiles used to initialize the model (after van den Heever *et al.*, 2006).

Table 8.2 *Aerosol initialization profiles for the sensitivity tests described in the text*

Experiment	Name	IN	CCN	GCCN
Exp1	EXP-CLN	Clean	Clean	Clean
Exp2	EXP-GCCN	Clean	Clean	Observed
Exp3	EXP-CCN	Clean	Observed	Clean
Exp4	EXP-IN	Observed	Clean	Clean
Exp5	EXP-CG	Clean	Observed	Observed
Exp6	EXP-IG	Observed	Clean	Observed
Exp7	EXP-IC	Observed	Observed	Clean
Exp8	EXP-OBS	Observed	Observed	Observed

The CLN and OBS vertical profiles of nucleating aerosol concentrations used to initialize the model are shown in Fig. 8.12. All of the model simulations were run for 12 hours from 1200 UTC to 0000 UTC (0800 LT to 2000 LT) on 28 July 2002. The simulations were initially run with three grids from 1200 UTC to 1800 UTC, after which the fourth grid was added. Eight sensitivity simulations were performed in which various combinations of the CLN and OBS initialization profiles of the three aerosol species were used. The sensitivity tests were identical in all other respects. In the control experiment (EXP-CLN), the clean profiles for all three of the aerosol species were utilized, while the OBS profiles for these three species were used for the EXP-OBS experiment. The combinations of the profiles used to initialize the model in the six other sensitivity tests are shown in Table 8.2.

In order to assess the individual and collective impacts of enhanced concentrations of CCN, GCCN, and IN on the surface precipitation produced by these storms, Alpert–Stein Factor Separation Methodology was performed using the output of all eight sensitivity experiments. The formulation of each factor pertaining to each

Table 8.3 *The components of the Alpert–Stein Methodology described in the text*

Factor	Description	Calculation
f_0	Part of the predicted field independent of the factors (clean air profiles)	$f_0 = \mathrm{Exp1}$
f_1	Part of the predicted field when only factor number 1 (IN) is fully switched on	$f_1 = \mathrm{Exp4} - \mathrm{Exp1}$
f_2	Part of the predicted field when only factor number 2 (CCN) is fully switched on	$f_2 = \mathrm{Exp3} - \mathrm{Exp1}$
f_3	Part of the predicted field when only factor number 3 (GCCN) is fully switched on	$f_3 = \mathrm{Exp2} - \mathrm{Exp1}$
f_{12}	Part of the predicted field dependent solely on the combination of factors 1 and 2 (IN and CCN)	$f_{12} = \mathrm{Exp7} - (\mathrm{Exp4} + \mathrm{Exp3}) + \mathrm{Exp1}$
f_{13}	Part of the predicted field dependent solely on the combination of factors 1 and 3 (IN and GCCN)	$f_{13} = \mathrm{Exp6} - (\mathrm{Exp4} + \mathrm{Exp2}) + \mathrm{Exp1}$
f_{23}	Part of the predicted field dependent solely on the combination of factors 2 and 3 (CCN and GCCN)	$f_{23} = \mathrm{Exp5} - (\mathrm{Exp3} + \mathrm{Exp2}) + \mathrm{Exp1}$
f_{123}	Part of the predicted field dependent solely on the combination of factors 1, 2 and 3 (IN, CCN, and GCCN)	$f_{123} = \mathrm{Exp8} - (\mathrm{Exp7} + \mathrm{Exp6} + \mathrm{Exp5}) + (\mathrm{Exp4} + \mathrm{Exp3} + \mathrm{Exp2}) - \mathrm{Exp1}$

of the nucleating species is shown in Table 8.3. The Alpert–Stein Methodology is extremely useful in that not only does it provide the impacts exerted by each nucleating species individually, but it also provides information about the impacts on surface precipitation that occur as a result of the presence of two of more of these nucleating aerosols. The surface precipitation analysis is performed for the time period between 1800 and 0000 UTC, the time period from when grid 4 is introduced to when the anvils start approaching the western boundary of grid 3.

8.2.3 Results

Storm development

Vertically integrated condensate for the EXP-OBS case is shown in Fig. 8.13. Convection (associated with regions of greater vertically integrated condensate) developed to the southwest of Lake Okeechobee and tracked toward the west and southwest, as is evident in the satellite imagery (Fig. 8.11). Extensive anvil cirrus clouds, evident in both the observations and the model output, develop. The

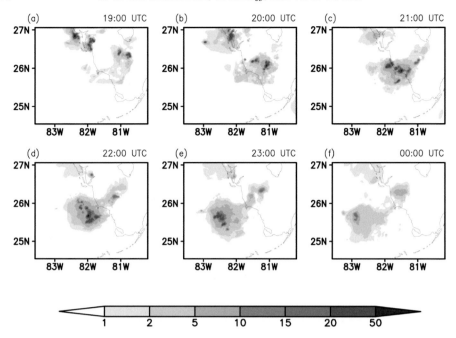

Figure 8.13 Vertically integrated total condensate (mm) at hourly intervals for the EXP-OBS simulation (after van den Heever *et al.*, 2006).

convection in the simulated output tends to move offshore more rapidly than was observed. This appears to be due to stronger cold pools in the simulations that overwhelm the sea-breeze circulation more rapidly, thus allowing the convective cells to move off the coast earlier than in the observations. However, in general, the observed storm development and associated cloud formation is well represented by the model output.

Accumulated surface precipitation

The accumulated surface precipitation (acre-feet) for the entire area of grid 3 for each of the sensitivity tests is shown in Table 8.4. It is apparent that by 1800 UTC most surface precipitation occurred in the experiments in which only the IN and GCCN concentrations were enhanced (EXP-IN, EXP-GCCN), followed by the case in which the IN and GCCN concentrations are increased simultaneously (EXP-IG). All three of these cases produce more surface precipitation than in the EXP-CLN experiment, thereby demonstrating an initial enhancement of the surface precipitation when increasing these nucleating aerosol concentrations. All of the sensitivity tests in which CCN concentrations were enhanced (EXP-CCN, EXP-CI, EXP-CG, EXP-OBS) result in a decrease in the surface precipitation at

Table 8.4 *Accumulated surface precipitation (acre-feet) for the eight sensitivity tests described in the text, in descending order, and the factors from the factor separation analysis, also in descending order, at 1800 and 0000 UTC*

1800 UTC				0000 UTC			
Accumulated precipitation		Factor separation		Accumulated precipitation		Factor separation	
Exp name	Magnitude (a-f)	Factor name	Contribution (a-f)	Exp name	Magnitude (a-f)	Factor name	Contribution (a-f)
EXP-IN	66 608	f_0 (CLN)	63 289	EXP-CLN	442 168	f_0 (CLN)	442 618
EXP-GCCN	65 874	f_{123} (I+C+G)	8 479	EXP-GCCN	368 053	f_{12} (I+C)	79 066
EXP-IG	63 487	f_1 (IN)	3 318	EXP-IG	352 112	f_{13} (I+G)	76 852
EXP-CLN	63 289	f_3 (GCCN)	2 584	EXP-IN	349 373	f_{23} (C+G)	57 335
EXP-CCN	61 741	f_2 (CCN)	−1 548	EXP-OBS	346 309	f_{123} (I+C+G)	−44 375
EXP-CG	58 275	f_{13} (I+G)	−5 705	EXP-CCN	344 338	f_3 (GCCN)	−74 114
EXP-IC	57 700	f_{23} (C+G)	−6 050	EXP-IC	330 610	f_1 (IN)	−92 794
EXP-OBS	57 008	f_{12} (I+C)	−7 359	EXP-CG	327 560	f_2 (CCN)	−97 829

this time, even when occurring with enhanced concentrations of GCCN and/or IN. These results for CCN and GCCN are generally consistent with the trends observed previously in the literature, in that CCN tend to suppress the formation of warm rain, while enhanced GCCN concentrations tend to enhance this process. The dominance of the sensitivity tests in which IN and GCCN concentrations are increased suggests a response similar to "dynamic seeding" concepts in which enhanced glaciation of convective clouds leads to dynamical invigoration of the clouds, larger amounts of processed water, and thereby enhanced rainfall at the ground (Simpson *et al.*, 1967; Rosenfeld and Woodley, 1989; 1993). This is described in more detail below.

By 0000 UTC the total surface precipitation is greatest in the CLN case, demonstrating the reduction in surface precipitation associated with the increases in aerosol concentrations, whether these aerosols are serving as CCN, GCCN, or IN. Of the dusty cases, increases in GCCN concentrations result in the most surface precipitation, followed by the enhancements of IN. All the simulations involving increases in CCN concentrations produce the least surface precipitation, even less than the case when all three nucleating species concentrations are enhanced. That the experiment when the concentrations of all three nucleating species are enhanced produces more precipitation than when CCN concentrations alone are enhanced suggests a nonlinear response to the presence of all three nucleating species.

Precipitation analysis using the Alpert–Stein Methodology

The Alpert–Stein Factor Separation Methodology (Stein and Alpert, 1993) was used to isolate the effects that each nucleating aerosol species (f_1, f_2, f_3) would have on the accumulated precipitation independent of enhancement in the concentrations of the other two aerosols, as well as the effects on surface precipitation that arise from the fact that one nucleating species is operating in the presence of one or both of the other two nucleating species (f_{12}, f_{13}, f_{23}, f_{123}). It should be noted here that the latter factors indicate the accumulated precipitation that arises as a result of one (or more) aerosol species being active in the presence of one (or more) of the other species, and are not simply the total precipitation output from the experiment in which the concentrations of two or more aerosols are enhanced. The latter case includes the precipitation produced by each nucleating aerosol species irrespective of the presence of one or more of the nucleating aerosols, as well as the precipitation produced as a result of the nucleating aerosols occurring simultaneously. For example, the factor f_{13} demonstrates the impact that the combined presence of GCCN and IN have on total precipitation, by subtracting the impacts that GCCN and IN have independently on the accumulated precipitation. The factor f_{13} should not therefore be confused with the total precipitation produced in EXP-IG.

Accumulated precipitation at 1800 UTC It is apparent from the Alpert–Stein Methodology shown in Table 8.4 that f_0 is greater than all the other factors, thus demonstrating that while increasing the concentrations of the various nucleating aerosols does influence the surface precipitation, the impacts are less significant than the other forces affecting surface precipitation, such as convection, updraft strength, and cold pool interactions. This remains true throughout the simulation as can be seen from the results at 0000 UTC. The factor analysis also demonstrates that enhanced concentrations of IN (f_1) or GCCN (f_3) result in an enhancement of the surface precipitation, while enhanced concentrations of CCN (f_2) cause a decrease in surface precipitation amounts. The impacts of the increased GCCN and IN concentrations have a more significant impact on the surface precipitation than for the CCN at 1800 UTC, as indicated by the absolute magnitude of these three factors (f_1, f_2, f_3). It is also evident from the factor analysis that, while enhanced concentrations of IN and GCCN increase the surface precipitation, as soon as these nucleating species are enhanced simultaneously (EXP-IG), or with CCN (EXP-CG, EXP-CI), the combined effect (f_{13}, f_{23}, f_{12}) is to decrease surface precipitation, with the experiments in which CCN is one of the two nucleating species (EXP-CG, EXP-IC) showing the greatest detrimental effect. Finally, it is interesting to note that, while the combined effect of increasing the concentrations of two of the species has a detrimental effect on surface precipitation, the combined effect of increasing the concentrations of all three species (f_{123}) not only enhances the surface precipitation, but this effect is the most significant of all the aerosol effects.

Accumulated precipitation at 0000 UTC By 0000 UTC, an interesting change in the influence of aerosol interactions has occurred. The Alpert–Stein Methodology indicates that at this time it is the combined effect of any two of the nucleating aerosol species (f_{12}, f_{13}, f_{23}) that has a positive impact on surface precipitation. Enhancing the concentrations of any of the nucleating aerosol species independently (f_1, f_2, f_3) is detrimental to surface precipitation totals, as is the combined effect of all three aerosols (f_{123}) when these aerosol concentrations are simultaneously increased. The enhanced CCN concentration has the greatest negative impact on surface precipitation at this stage, closely followed by the enhanced concentrations in IN. However, the combined influence of these two species results in a contribution that is approximately 80–85% of that of each individual contribution, and this combined contribution to the surface rainfall is positive.

The question that needs to be asked here is why the combined effect between any of the two aerosols has a positive contribution to the accumulated surface precipitation, whereas the individual impact of each of these nucleating aerosols is negative? The following scenarios could help to explain some of these observations at 0000

UTC. Consider the EXP-IC experiment – both the Alpert–Stein Methodology and the precipitation totals demonstrate that enhanced CCN concentrations reduce the surface precipitation. This occurs as the greater number concentrations of CCN will result in the formation of a greater number of smaller cloud droplets, which are less efficient at colliding and coalescing. This results in larger amounts of cloud water being present in the cloud that can then be lofted vertically by the updrafts, where it will freeze either by riming and raindrop collection of ice crystals or activation of IN. The greater concentrations of IN provide greater opportunities for freezing, which in turn results in greater amounts of latent heat being released, thereby enhancing the strength of the updrafts, the intensity of the storms, and the surface precipitation produced. Stronger and more numerous updrafts are observed in this experiment (not shown here). Thus, the simultaneous enhancements in CCN and IN concentrations work collectively to enhance the surface precipitation in that the CCN increase the amount of cloud water available for vertical transport and subsequent freezing, and the IN provide added sites for ice formation and subsequent latent heat release.

In the case when CCN and GCCN are simultaneously enhanced (EXP-CG), the enhanced CCN concentrations once again result in a greater number of small cloud droplets. These smaller cloud droplets are then efficiently collected by the larger cloud droplets that develop as a result of the formation of precipitation embryos by the enhanced GCCN concentrations. Thus these two species work together in that the enhanced CCN concentrations result in greater amounts of cloud water being available for the formation of raindrops, and the enhanced GCCN concentrations generate greater numbers of larger cloud droplets which collect the smaller cloud droplets generated by the CCN.

Finally, it should be pointed out that the simulations are simply initialized with the dusty vertical profiles, which is certainly representative of some observed scenarios. However, situations will also arise in which dust is continuously transported into a region. That situation is not represented here. Given that these experiments are not continuously updated with an aerosol source, the various aerosols do get washed out from the atmosphere, a factor that needs to be considered when comparing the results between 1800 UTC and 0000 UTC.

8.2.4 Conclusions

The application of the Alpert–Stein Methodology to the results of the simulations presented in this paper demonstrates that aerosol variations can have a significant impact on accumulated surface precipitation. The accumulated surface precipitation is initially greater in the cases in which the GCCN and/or IN concentrations are enhanced than in the CLN run. However, at the end of the simulation, while enhancements in both GCCN and IN still produce more precipitation at the surface than when CCN concentrations are increased, the accumulated precipitation is greatest in the CLN case, thus demonstrating a reduction in surface precipitation associated with increases in aerosol concentrations. It should be pointed out though, that by the end of the simulation the aerosol concentrations have been significantly reduced due to rainout, and hence their effects on surface precipitation are less significant.

The results shown here, obtained through the use of applying the Alpert–Stein Factor Separation Methodology to the output of a cloud-resolving numerical model, highlight the fact that the impacts of varying GCCN and IN concentration are just as significant as those associated with CCN, and that they need to be considered when examining the impacts of aerosols, such as Saharan dust, on convective storm characteristics. Perhaps even more importantly, the factor analysis clearly demonstrates the need to take into account the combined influence of these nucleating species when considering the impacts of Saharan dust on surface precipitation, a research task that has seldom been performed until now.

9

Free and forced thermocline oscillations in Lake Tanganyika

O. Gourgue, E. Deleersnijder, V. Legat, E. Marchal, and L. White

All year long, the thermocline of Lake Tanganyika (Central Africa) oscillates about two equilibrium states. The thermocline is tilted downward toward the north during the dry season, due to the wind bringing the warm surface water from south to north. The equilibrium position of the thermocline is horizontal during the wet season. The oscillations about these two equilibrium states may be of two types. The free oscillations are due to the seasonal cycle of the wind stress, while the forced oscillations are a direct response to the intraseasonal variability of the surface forcing. It has already been suggested that both have a three- to four- week oscillation period. The Factor Separation method is here used to show that the forced oscillations of the thermocline are about twice as large as the free ones.

9.1 Introduction

Lake Tanganyika is located to the east of central Africa, and is shared by four developing countries: Democratic Republic of the Congo, Burundi, Tanzania, and Zambia. It lies between $3°20'$ and $8°45'$ S and $29°05'$ to $31°15'$ E. It is about 650 km long and 50 km wide on average. The mean depth of the lake is about 570 m, with a maximum depth of 1470 m (Fig. 9.1). That makes it the second deepest lake in the world, the deepest being Lake Baikal in Russia. Thermal stratification is well marked and present all year long, so that one can identify two distinct layers: the surface and the bottom layers. The surface layer is composed of relatively warm water (24–28 °C) while the bottom layer is composed of cooler water ($\sim$23.5 °C). These two layers are separated by a thermocline, which is a thin layer where the temperature gradient is maximum. The mean depth of the thermocline is about

Factor Separation in the Atmosphere: Applications and Future Prospects, ed. Pinhas Alpert and Tatiana Sholokhman. Published by Cambridge University Press. © Cambridge University Press 2011.

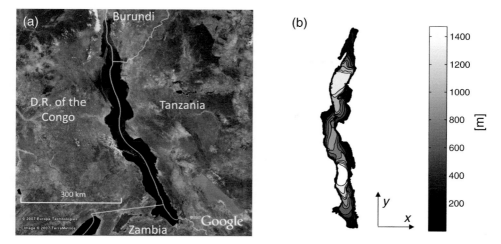

Figure 9.1 (a) Aerial view of Lake Tanganyika (from Google Earth^{TM} mapping service), indicating the four neighboring countries. (b) Bathymetry of Lake Tanganyika (with depths measured in meters) and x- and y-axis of the Cartesian coordinate system used in the present work.

50 m. As the very deep water almost never reaches the surface, the lake can be categorized as meromictic (Coulter and Spigel, 1991).

This study only focuses on the hydrodynamics. However, it must be kept in mind that the hydrodynamics exerts a profound influence on a number of chemical (Plisnier *et al.*, 1999) and biological (Descy *et al.*, 2005) properties throughout the lake. It also governs upwelling of nutrients that is essential in maintaining the pelagic food web (O'Reilly *et al.*, 2003). Water motions are chiefly forced by the surface wind stress. The region undergoes two seasons: the dry season (approximately from May to August), characterized by strong winds mainly blowing northwestward along the main axis of the lake, and the wet season (approximately from September to April), during which the winds are generally weaker.

During the dry season, the wind pushes the warmer surface water toward the northern end of the lake. The thermocline moves upward to compensate for the loss of water at the south and the warmer water accumulated at the north is pulled downward by gravity. The thermocline is then tilted downward toward the north and occasionally outcrops in the southern part of the lake. At the end of the dry season, when the southeasterly winds stop, the surface and bottom layers slide over each other, and the thermocline oscillates to reach a new equilibrium. These waves are reflected at the lake boundaries and gradually transform into standing wave patterns, called internal seiches. By December, the thermocline reaches its mean level (around 50 m) but keeps oscillating until the beginning of the next dry season and the onset of the southeasterly winds. The period of these oscillations is about three to four weeks. Indeed, a fundamental oscillation mode of 25 to 30 days has been found both from field data (Coulter and Spigel, 1991) and from a simple model (Mortimer, 1974), and a fundamental mode of about 24 days has been found from a reduced-gravity model with a seasonal wind stress forcing (Naithani *et al.*, 2003). The presence of rather small internal Kelvin waves in the lake has been emphasized by Naithani and Deleersnijder (2004) and Antenucci (2005).

Herein two types of oscillations are identified: the free and the forced ones. The free oscillations are related to the seasonal cycle of the surface wind stress. When a seasonal wind stress is applied over the whole lake (i.e., a constant wind stress during the dry season and zero wind stress during the wet season, see Fig. 9.2b), the thermocline oscillates about two equilibrium states: a tilted thermocline (downward toward the north) during the dry season and a horizontal one during the wet season (Naithani *et al.*, 2003). The free oscillations are not a direct response to the seasonal wind stress. They are due to the alternation between two different equilibrium states along the year. Therefore, free oscillations can be regarded as triggered by the seasonal wind stress. The period of the first mode of these free oscillations (which is the only significant mode) is about three to four weeks (Naithani *et al.*, 2003).

The forced oscillations are due to the intraseasonal variability of the surface wind stress, which is defined as the difference between the actual wind stress (Fig. 9.2a) and the seasonal wind stress (Fig. 9.2b). By means of a wavelet analysis, it has been shown that, in the region of Lake Tanganyika, this intraseasonal variability of the surface forcing has a period of three to four weeks (Naithani *et al.*, 2002). This period is due to the eastward-propagating low-frequency large-scale convection and circulation cells, which get their energy from the seasonally migrating intertropical convergence zone (Madden and Julian, 1971, 1994), and also to cloud-radiation (Krishnamurti and Bhalme, 1976) and evaporation-wind (Neelin *et al.*, 1987; Lin *et al.*, 2000) feedback processes, and to the interactions between moist convective and dynamical processes (Goswami and Shukla, 1984). The timescale associated with the intraseasonal variations of the wind forcing is of the same order of magnitude as the period of the free oscillations. Therefore, the oscillations directly forced by the intraseasonal oscillations of the wind stress tend to have a large amplitude, as was seen by Naithani *et al.* (2002) who identified this near-resonance phenomenon.

Free (Naithani *et al.*, 2003) and forced (Naithani *et al.*, 2002) thermocline oscillations in Lake Tanganyika have already been studied separately. However, so far, the relative importance of these two types of response and the interactions between them have not been investigated. Doing so by means of the Alpert–Stein Factor Separation Methodology is the objective of the present chapter. The wind stress is split into a seasonal component and an intraseasonal one. This leads to a decomposition of the thermocline response into a seasonal and an intraseasonal response

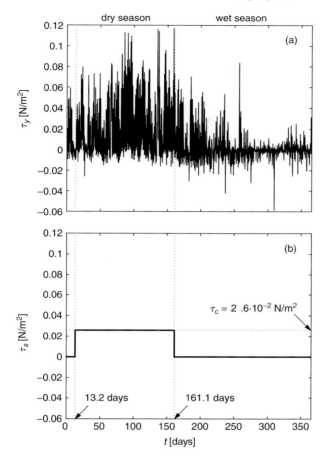

Figure 9.2 (a) Wind stress in the along-lake direction, measured at Mpulungu in Zambia, every six hours from 1 April 1993 to 31 March 1994, during the FAO/FINNIDA project "Research for the Management of the Fisheries on Lake Tanganyika" GCP/RAF/271/FIN. (b) Seasonal wind stress, with a constant wind stress τ_c during the dry season and zero wind stress during the wet season.

(respectively responsible for the free and forced oscillations) and also a synergistic term, due to the nonlinearity of the problem (Stein and Alpert, 1993). Therefore, the Alpert–Stein Methodology can help in assessing the relative magnitude of the free and forced oscillations of the thermocline in Lake Tanganyika.

9.2 Model description

Two-layer models have often been used to simulate wind-induced internal waves in large, stratified lakes. For example, we may cite studies of Lake Biwa in Japan

(Kanari, 1975), Lake Ontario in Canada (Schwab, 1977), or Lake Kinneret in Israel (Antenucci and Imberger, 2001). The equations of such models are well known, and references on their design may easily be found in the literature (Csanady, 1967; Kielmann and Simons, 1984). With two-layer models, there are six variables, i.e., the displacement of the free surface, the displacement of the interface between the two layers, the horizontal velocity components averaged over the surface-layer thickness and the horizontal velocity components averaged over the bottom-layer thickness.

When the bottom layer is much thicker than the surface one, relevant simplifications may lead to the reduced-gravity model (Naithani *et al.*, 2003). The reduced-gravity model is a 1.5-layer model in that it focuses on the dynamics of the surface layer, without disregarding completely the bottom one. It has only three variables, i.e., the displacement of the interface between the two layers and the horizontal velocity components averaged over the surface-layer thickness. The bathymetry is no longer explicitly taken into account.

Several papers have already shown that a reduced-gravity model is able to simulate rather well the thermocline oscillations in Lake Tanganyika (Naithani *et al.*, 2002, 2003) since stratification is present all year round and the surface layer is much shallower than the bottom one, i.e., the thickness of the former is about 10% of that of the latter (Coulter and Spigel, 1991).

The equations of the reduced-gravity model read:

$$\frac{\partial \xi}{\partial t} + \nabla \cdot (H\mathbf{u}) = 0 \tag{9.1}$$

$$\frac{\partial}{\partial t}(H\mathbf{u}) + \nabla \cdot (H\mathbf{u}\mathbf{u}) = -f\mathbf{e_z} \times (H\mathbf{u}) - (\varepsilon g)H(\nabla \xi) + \nabla \cdot (H\nu(\nabla \mathbf{u})) + \frac{\tau}{\rho} \tag{9.2}$$

where t denotes the time and ∇ is the horizontal spatial derivative vector operator; ξ is the downward displacement of the thermocline (Fig. 9.3), while $\mathbf{u}$ is the horizontal velocity vector averaged over the surface-layer thickness $H = h + \xi$ (since the displacement of the free surface is neglected), h being the surface-layer thickness at rest (Fig. 9.3); f is the Coriolis parameter; (εg) is the reduced gravity, where g is the gravitational acceleration and $\varepsilon = \frac{\rho_b - \rho}{\rho_b}$ is the relative density difference (here equal to 6.3×10^{-4} in order to be consistent with former studies); ν is the horizontal eddy viscosity; τ is the surface wind stress vector; ρ and ρ_b are the constant water densities in the surface layer an the bottom layer, respectively.

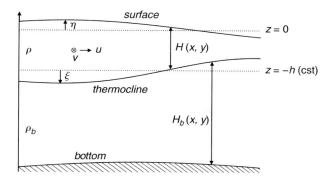

Figure 9.3 The parameters and variables of a reduced-gravity model: η is the upward displacement of the free surface, ξ is the downward displacement of the thermocline; ρ and ρ_b are the densities of the surface and bottom layers, respectively; u and v are the components of $\mathbf{u}$, the velocity vector averaged over the surface layer; h is the surface layer thickness at rest; H and H_b are the instantaneous surface- and bottom-layer thicknesses, respectively.

The standard reduced-gravity model rests on the assumptions that the thermocline is impermeable, implying that the volume of the surface layer remains constant. However, some modifications allow us to take into account the water exchanges between the surface and bottom layers, and those due to precipitation, evaporation and rivers (Gourgue *et al.*, 2007). This approach is not resorted to herein, in order to follow the assumptions of the previous studies (Naithani *et al.*, 2002, 2003), and because it is not essential for studying the internal oscillations in the lake.

Since the rivers are not taken into account, the lateral boundary of the lake can be assumed to be impermeable. Moreover, we make the assumption of a free slip lateral boundary. The corresponding boundary conditions therefore read:

$$u_n = 0 \tag{9.3}$$

$$\frac{\partial u_t}{\partial n} = 0 \tag{9.4}$$

where u_n and u_t denote the horizontal velocity components normal and tangential to the boundary, respectively, and $\frac{\partial}{\partial n}$ the spatial derivative in the direction normal to the boundary.

To simulate the oscillations of the thermocline, we discretize Eqs. (9.1) and (9.2) on the unstructured mesh displayed in Fig. 9.4, using the two-dimensional component of the finite element model SLIM (Second-generation Louvain-la-Neuve

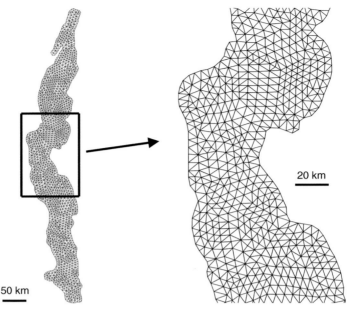

Figure 9.4 Unstructured mesh of Lake Tanganyika used for the present simulations. There are 2997 triangular elements, whose mean size is about 5 km.

Ice-ocean Model[1]). The elevation and velocity variables are approximated by linear conforming (P_1) and linear nonconforming (P_1^{NC}) shape functions, respectively.
Therefore, the elevation nodes are situated at the vertices of each triangle of the
mesh, and the velocity nodes at the middle of their edges (Hanert *et al.*, 2005).

Occasionally, the upward displacement of the thermocline can be equal to the
reference thickness of the surface layer. To prevent the thermocline from outcropping, a wetting–drying algorithm is implemented. When the water column depth of
a node reaches a critical value, which is taken to be 5 m in this application, the node
is considered to be dry. Otherwise, it is a wet node. The wetting–drying algorithm
consists in setting to zero the fluxes across the edges of a patch made up of all
the triangles sharing a dry node. At each time step, we first compute the elevation
field ξ. If there are dry nodes, we modify the velocity fields u and v (of the previous
time step) with the wetting–drying algorithm, before recomputing the elevation
field. This technique is used recursively until there are no more dry nodes. Then,
we compute simultaneously the velocity fields u and v. This kind of approach was
initially introduced for the finite difference method (Balzano, 1998). It has already

---

[1] www.climate.be/SLIM

been used in a finite difference model of Lake Tanganyika (Naithani *et al.*, 2002, 2003), and it has been adapted here to our finite element model (Marchal, 2005).

The initial conditions need no discussion: the numerical results are analyzed several years after the beginning of the simulation, i.e., at a time when a periodic regime is reached that is independent of the initial values of the variables.

9.3 Wind stress forcing decomposition

Surface wind measurements were made from 1 April 1993 to 31 March 1994 at Mpulungu by the FAO/FINNIDA project "Research for the Management of the Fisheries on Lake Tanganyika" GCP/RAF/271/FIN. Mpulungu is situated at the south end of the lake in Zambia. Following former studies, we assume that this pointwise wind stress may be applied to the entire lake surface (Naithani *et al.*, 2002). Therefore, the surface wind stress vector $\tau(t)$ is a function of time only. Its components are $\tau_x(t)$ and $\tau_y(t)$, following the *x*- and *y*-axis of Fig. 9.1b, respectively (the *y*-axis is in the along-lake direction, from south to north). The component $\tau_x(t)$ of the surface forcing may be neglected, for it is much smaller than $\tau_y(t)$, and because of the narrowness of the lake (Naithani *et al.*, 2002). Accordingly, $\tau_y(t)$ is assumed to be the actual wind stress (Fig. 9.2a).

In order to evaluate the relative importance of the free and forced thermocline responses, the actual wind stress is split into two parts:

$$\tau_y(t) = \tau_s(t) + \tau_i(t) \tag{9.5}$$

where $\tau_s(t)$ is the seasonal wind stress (triggering the free oscillations), and $\tau_i(t)$ is the intraseasonal variability of the wind stress (causing the forced oscillations). The seasonal wind stress is made up of a constant wind stress τ_c during the dry season, and zero wind stress during the wet season, as shown in Fig. 9.2b. It is thus defined by three parameters, the beginning and the duration of the dry season, and the value of the constant wind stress during the dry season τ_c. The values of these parameters (given in Fig. 9.2b) are determined by minimizing the integral of the least-square difference between the functions $\tau_y(t)$ and $\tau_s(t)$.

9.4 Thermocline response decomposition

The downward displacement of the thermocline, $\xi(t, x, y)$, is defined as the thermocline response to the actual wind stress forcing $\tau_y(t)$. We call it the complete thermocline response, and it can be decomposed as follows:

$$\xi(t,x,y) = \underbrace{\xi_{sno}(t,x,y) + \xi_{so}(t,x,y)}_{\xi_s(t,x,y)} + \xi_i(t,x,y) + \xi_{syn}(t,x,y) \qquad (9.6)$$

where

- the seasonal response $\xi_s(t,x,y)$ is the thermocline response to the seasonal wind stress $\tau_s(t)$;
- the intraseasonal response $\xi_i(t,x,y)$ is the thermocline response to the intraseasonal variability of the wind stress $\tau_i(t)$;
- the synergistic term $\xi_{syn}(t,x,y)$ is due to the nonlinearity of the model (Stein and Alpert, 1993).

The seasonal response ξ_s cannot be considered as purely oscillatory. Indeed, the thermocline oscillates about two equilibrium positions: a tilted thermocline (downward toward the north) during the dry season, and a horizontal one during the wet season, as is explained in the appendix at the end of this chapter. Therefore the seasonal response ξ_s is split into two parts:

- the reference nonoscillating part of the seasonal response $\xi_{sno}(t,x,y)$ is made up of two equilibrium states: it is the time-independent response to τ_c during the dry season, and it is zero during the wet season;
- the oscillating part of the seasonal response $\xi_{so}(t,x,y)$ represents the oscillations of the seasonal response about these two equilibrium states.

The components ξ, ξ_s, and ξ_i are computed by means of the reduced-gravity model, using τ_y, τ_s, and τ_i, respectively, as wind stress forcing. The dry season part of ξ_{sno} is obtained by means of the reduced-gravity model using τ_c as wind stress, and its wet season part is zero. The other components ξ_{syn} and ξ_{so} are deduced from them since $\xi_{syn} = \xi - \xi_s - \xi_i$ and $\xi_{so} = \xi_s - \xi_{sno}$.

Only two of these contributions may be regarded as oscillations: ξ_{so} and ξ_i are now defined as the free and forced thermocline oscillations, respectively. For the reader's convenience, ξ_{sno} will be termed the nonoscillating seasonal response. The names and symbols of all the thermocline responses are gathered in Table 9.1. We now define several tools to analyze all these contributions.

9.4.1 Width-average

The thermocline responses can be averaged over the lake width:

$$\overline{\xi}(t,y) = \frac{\int_{w_1(y)}^{w_2(y)} \xi(t,x,y)dx}{\int_{w_1(y)}^{w_2(y)} dx}. \qquad (9.7)$$

Table 9.1 *Meaning of the symbols associated with the various components of the thermocline reponse to the wind forcing*

ξ	total response
ξ_s	seasonal response
ξ_{sno}	nonoscillating seasonal response
ξ_{so}	free oscillations (i.e., oscillating seasonal response)
ξ_i	forced oscillations (i.e., intraseasonal response)
ξ_{syn}	synergistic term

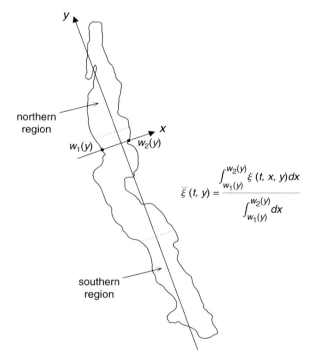

$$\bar{\xi}(t, y) = \frac{\int_{w_1(y)}^{w_2(y)} \xi(t, x, y)\,dx}{\int_{w_1(y)}^{w_2(y)} dx}$$

Figure 9.5 A width-averaged response $\bar{\xi}(t, y)$ is computed over the lake width that depends on y. This width is determined by $w_1(y)$ on the left and $w_2(y)$ on the right. Responses averaged over the southern and northern regions of the lake are averaged over the regions of the lowest third and top two-thirds of the lake length, respectively.

The lake width is determined by $w_1(y)$ on the left and $w_2(y)$ on the right (Fig. 9.5). Figure 9.6 presents the width-averaged annual cycles of the complete thermocline response ξ and its four contributions: the non-oscillating seasonal response ξ_{sno}, the free oscillations ξ_{so}, the forced oscillations ξ_i, and the synergistic term ξ_{syn}.

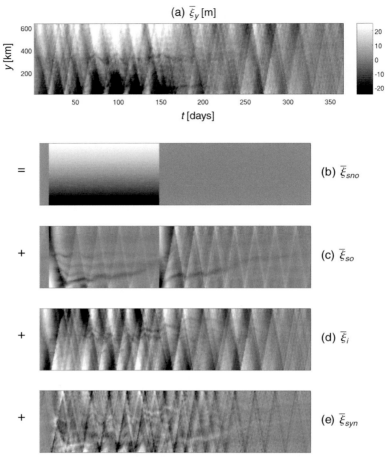

Figure 9.6 Annual cycles of the width-averaged complete response (a), the width-averaged nonoscillating seasonal response (b), the width-averaged free oscillations (c), the width-averaged forced oscillations (d), and the width-averaged synergistic term (e).

The complete response is clearly oscillating, and the thermocline is more tilted downward toward the north during the dry season, which is why we have introduced ξ_{sno}. The free oscillations seem to be only significant at the season changes. This is not the case with the forced oscillations that are present all year long, even if they are larger during the dry season. It has to be noted here that the relatively quick damping of the free oscillations is mainly due to the viscosity ν, which is chosen to be equal to $3\ \mathrm{m}^2\ \mathrm{s}^{-1}$ here in order to be consistent with the former studies (Naithani *et al.*, 2002, 2003). The synergistic term is not negligible and is also of an oscillating nature.

9.4.2 *Average over a specific region*

The thermocline responses can be averaged over a specific region Ω_i:

$$\hat{\xi}(t) = \frac{\int_{\Omega_i} \xi(t,x,y)dxdy}{\int_{\Omega_i} dxdy}. \tag{9.8}$$

We define a southern and a northern region (Fig. 9.5). Figure 9.7 presents the mean annual cycles of four thermocline responses (ξ, ξ_{sno}, ξ_{so}, and ξ_i) in these two regions. In the south (north), the oscillating responses present 13 local minima (maxima)

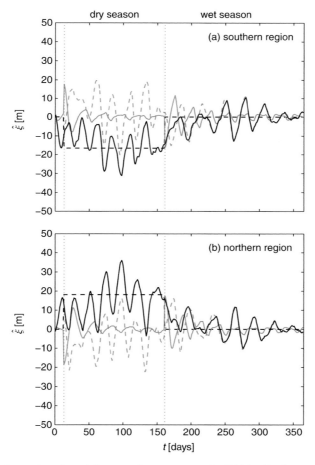

Figure 9.7 Annual cycles of the complete response (solid black), the nonoscillating seasonal response (dashed black), the free oscillations (solid gray) and the forced oscillations (dashed gray), averaged over the southern (a) and northern (b) regions of the lake. These regions are defined as the lower one-third and the top two-thirds of the lake length, respectively (Fig. 9.5).

per year, corresponding to an oscillation period of about 28 days, which is consistent with the above-mentioned three- to four-week oscillation period. Figure 9.7 also shows that the amplitude of the forced oscillations is much larger than that of the free ones (which are only significant at the season changes, as mentioned previously). Moreover the forced oscillations are more likely to fit the complete response than the free ones. The complete thermocline response ξ seems to be essentially made up of both the nonoscillating seasonal response ξ_{sno} and the intraseasonal response ξ_i.

9.4.3 *Average over the whole lake: the concept of mean amplitude*

Since the surface layer volume is constant, the average of a thermocline response over the whole lake is zero. We then introduce the concept of mean amplitude, which uses the absolute value of the thermocline responses. We define the mean amplitude of a thermocline response as follows:

$$\tilde{\xi}(t) = \frac{\int_\Omega |\xi(t,x,y)| \, dxdy}{\int_\Omega dxdy}, \qquad (9.9)$$

where Ω is the whole lake surface. The annual mean amplitudes (i.e., $\tilde{\xi}$ averaged over the whole year) of the free and forced oscillations are equal to 7.87 m and 18.75 m, respectively. The mean amplitude of the forced oscillations is thus, on average, more than twice as large as that of the free ones.

Figure 9.8 presents the evolution of the mean amplitude of the free and forced oscillations and their ratio. The forced oscillations are present all year round. Their mean amplitude is maximum during the dry season and then decreases during the wet season. Their amplitude is always larger than that of the free oscillations, except at the season changes when the amplitude of the free oscillations strongly increases before decreasing exponentially. Except at the season changes, the mean amplitude of the forced oscillations is over twice as large as that of the free oscillations, and even more so during the dry season (about five times on average).

The annual mean amplitude of the synergistic term is equal to 13.40 m. Therefore, the synergistic term ξ_{syn} is not negligible. This is partly due to the nonlinearity of the reduced-gravity model equations. The wetting–drying algorithm, which prevents the thermocline from outcropping, also adds some nonlinearity to the model. Moreover, the seasonal and intraseasonal responses are often out of phase, which is likely to increase the amplitude of the synergistic term.

9.4.4 *Selecting the most realistic reponse components*

Is the complete thermocline response mainly composed of the nonoscillating seasonal response ξ_{sno}, the free oscillations ξ_{so}, the forced oscillations ξ_i, or a

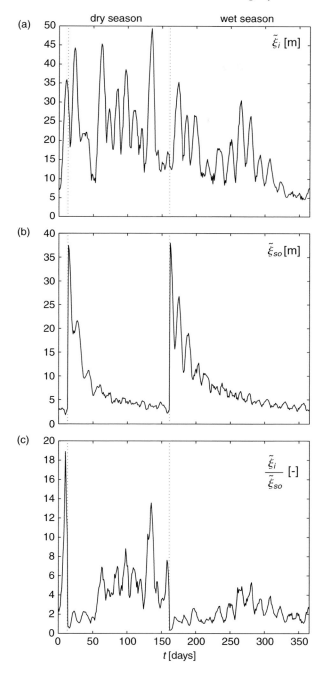

Figure 9.8 Evolution of the mean amplitude [m] of the forced (a) and free (b) oscillations, and evolution of their ratio [–] (c).

Table 9.2 *Mean amplitude of the*
different signals composed of the
difference between ξ and the various
combinations of ξ_{sno}, ξ_{so}, *and* ξ_i

Signal	Amplitude
$\xi - (\xi_{sno} + \xi_i)$	13.12 m
$\xi - (\xi_{sno} + \xi_{so} + \xi_i)$	13.4 m
$\xi - \xi_{sno}$	16.75 m
$\xi - (\xi_{sno} + \xi_{so})$	18.39 m
$\xi - \xi_i$	22.74 m
$\xi - \xi_{so}$	23.27 m
$\xi - (\xi_{so} + \xi_i)$	23.43 m

combination of several of these three thermocline responses? There are seven possible combinations, composed of one, two or three terms taken among ξ_{sno}, ξ_{so}, and ξ_i. We call a signal the difference between the complete response ξ and one of these combinations. That the mean amplitude of a given signal is small indicates that the associated combination is a good approximation of the complete response. Table 9.2 presents the mean amplitude of the signals related to the seven possible combinations. It shows that the four combinations containing ξ_{sno} are the best ones. So the nonoscillating seasonal response is essential to describe the complete response. According to this method, the best combination is $\xi_{sno} + \xi_i$. The free oscillations ξ_{so} are not only useless to describe the complete response, but the approximation to the complete response gets worse when ξ_{so} is added.

9.5 Conclusion

A hydrodynamic model was built that reproduces rather well the results of former simulations of the thermocline oscillations in Lake Tanganyika. Former studies established the presence of free oscillations with a period of three to four weeks (Naithani *et al.*, 2003), and noted that the thermocline oscillations are also due to forced oscillations with the same three- to four-week period (Naithani *et al.*, 2002) (the period of these forced oscillations is due to the intraseasonal variability of the wind stress in the region of the lake). But no comparison between the two oscillation types had been achieved. We showed that, in Lake Tanganyika, the forced oscillations of the thermocline are significantly larger than their free counterparts: their mean annual amplitude is more than twice as large. It was also seen that combining the nonoscillating seasonal response of the thermocline (which is tilted downward toward the north during the dry season and horizontal for the rest of the year) and the forced oscillations yields a response that is closer to the complete one than any other response combination.

Appendix Seasonal response decomposition

As illustrated by Fig. 9.2b, the seasonal component of the surface wind stress is defined to be

$$\tau_s(t) = \begin{cases} \tau_c & \text{during the dry season} \\ 0 & \text{during the wet season} \end{cases} \tag{9.10}$$

where τ_c is a positive constant. The response of the thermocline to this wind forcing, i.e., the seasonal thermocline response ξ_s, is made up of an oscillating part ξ_{so} and a nonoscillating one ξ_{sno}. The latter is the steady-state response to the seasonal wind stress. Accordingly, it obeys the momentum equation

$$\varepsilon g(h + \xi_{sno})\frac{\partial \xi_{sno}}{\partial y} = \frac{\tau_s}{\rho} \tag{9.11}$$

Then, it is readily seen that the seasonal nonoscillating thermocline displacement is zero during the wet season (i.e., the thermocline is horizontal) and is equal to the following expression during the dry season:

$$\xi_{sno}(t, x, y) = -h + \sqrt{\frac{2\tau_c(y + Y)}{\varepsilon g \rho}} \tag{9.12}$$

where Y is a constant whose value must guarantee that the volume of the surface layer remains time-independent. In other words, this constant is such that the integral of ξ_{sno} over the surface of the lake Ω is zero, implying that Y must satisfy

$$\int_\Omega h\, dx\, dy = \int_\Omega \sqrt{\frac{2\tau_c(y + Y)}{\varepsilon g \rho}}\, dx\, dy \tag{9.13}$$

Acknowledgements

Eric Deleersnijder and Laurent White are a Research Associate and an honorary Postdoctoral Researcher, respectively, with the Belgian National Fund for Scientic Research (FNRS). The present study was carried out within the scope of the project "A Second-Generation Model of the Ocean System", which is funded by the Communauté Française de Belgique, as Actions de Recherche Concertées, under contract ARC 04/09–316. This work is a contribution to the construction of SLIM, the Second-generation Louvain-la-Neuve Ice-ocean Model (www.climate.be/SLIM).

The authors are indebted to Jaya Naithani and Pierre-Denis Plisnier for useful discussions. Finally, the present study is linked to the CLIMFISH project (Climate Change Impact on the Sustainable Use of Lake Tanganyika Fisheries), funded by the STEREO programme of Belgian Federal Science Policy and the framework agreement between DGCD (Belgian Development Cooperation) and the Royal Museum for Central Africa (RMCA), Tervuren, Belgium.

10

Application of the Factor Separation Methodology to quantify the effect of waste heat, vapor and pollution on cumulus convection

G. W. Reuter

Large oil refineries emit heat, vapor, and cloud condensation nuclei (CCN), all of which can affect the formation of cloud and precipitation. The Factor Separation (FS) technique is applied to isolate the net contributions of waste heat, vapor, and CCN to the rainfall of a cumulus developing in the industrial plume. The mutual-interactive contributions of two or three of the factors are also computed. The model simulations indicate that the sensible heat provides the major stimulus for cloud development and rain formation. The pure contribution of the industrial CCN is to enhance the condensation causing an increase in the mass of total cloud water. The contributions arising from mutual interactions among two or three factors are quite large and should not be neglected. Particularly, the synergistic interaction of the sensible heat and pollution effects contribute towards the accumulated rainfall.

10.1 Introduction

There is considerable interest in the effects of large electrical power plants and oil refineries on meteorological phenomena. Preferential cumulus formation has been observed above electrical power plants and oil refineries (Auer, 1976). Hobbs *et al.* (1970) reported that in regions adjacent to or downwind of the Port Townsend paper mill (Washington State, USA) the annual rainfall recorded was 30% greater than the rainfall from nearby stations. This dramatic increase in annual precipitation is likely caused by the presence of the paper mill. Hobbs *et al.* speculated that the enhanced rainfall might be attributed to the large and giant CCN emitted from the paper mill into the pollution plume. Support for this hypothesis came from Eagan *et al.*'s (1974) study. They found a much broader spectrum of cloud droplet sizes in clouds

Factor Separation in the Atmosphere: Applications and Future Prospects, ed. Pinhas Alpert and Tatiana Sholokhman. Published by Cambridge University Press. © Cambridge University Press 2011.

that formed in the pollution plume over the paper mill than in clouds that formed in the ambient air unaffected by the plume. In addition, the polluted clouds contained a significant number of droplets with diameters larger than 30 μm, which were rare in the unpolluted clouds. Thus, polluted clouds were believed to have a more efficient collision–coalescence mechanism of raindrop formation. However, Hindman *et al.* (1977b) made calculations of the growth of droplet distributions using the observed spectrum from the paper mill as input data. The model results indicated that the higher concentrations of large droplets in the pollution plume from the paper mill did not cause a significant enhancement in rainfall. They concluded that the large and giant CCN emitted by the mill cannot by themselves be responsible for the observed increase in local rainfall. Instead, they suggested that sensible heat and vapor emitted by the mill, in combination with the pollution, accounted for the observed rain stimulation.

Similar findings were reported for the effects on rainfall near a large oil refinery located at Wood River, St. Louis, USA. Rainfall measurements taken from just upwind and downwind of the refinery indicated that the refinery caused an increase in precipitation of 38% during the three summer months. Modelling studies for a particular case showed that the waste energy in the form of sensible and latent heat provided a large stimulus to the triggering and development of cumulus convection and rain showers (Murray *et al.*, 1978; Guan and Reuter, 1995). On the other hand, the impact of the pollution for this case was rather small (Reuter and Guan, 1995).

Large industrial complexes emit waste energy as sensible heat and vapor, as well as industrial aerosols that act as efficient CCN. Experts agree that each of these three factors can contribute to cloud formation. However, the relative importance of the three factors and their possible interactions are not obvious. The purpose of this chapter is to clarify the role of waste heat, vapor, and pollution on rain enhancement for short-lived cumuli developing in an industrial plume. We here focus on calm wind conditions that permits the use of an axisymmetric cumulus model.

To quantify the relative importance of the three factors, we have to isolate their individual contributions, as well as the contributions due to their mutual interactions. The Factor Separation Methodology suggested by Stein and Alpert (1993) can isolate these individual and mutual contributions and will thus be used.

10.2 Model framework

The numerical cloud model used in this study is an extension of the time-dependent cumulus model developed by Reuter and Yau (1987) and Guan and Reuter (1995). The model code was extended to deal with warm rain microphysics in polluted and unpolluted conditions using a bulk-water parameterization scheme

(Reuter and Guan, 1995). Only the major characteristics of this model are briefly summarized here.

The anelastic set of equations written in cylindrical coordinates is used. Assuming axisymmetry, the azimuthal gradients of all quantities are set to zero. The subgrid-scale processes are modelled using a first-order closure scheme in which the eddy exchange coefficient depends on the local deformation shear and the local moist buoyancy time scales. Bulk parameterization of cloud and rain is used. The liquid water is divided into cloud water (consisting of small droplets moving with the air) and rain water (consisting of larger drops falling downwards). Following Berry and Reinhardt (1973), the rate of autoconversion of cloud water to rain water depends on the cloud water mixing ratio, the initial cloud droplet concentration, and the mass dispersion coefficient of the initial cloud droplet distribution. Since the cloud droplet size spectrum depends on the distribution of activated CCN, the autoconversion rate into rain becomes dependent on the pollution level of the air. The supersaturation value required for cloud condensation is also related to the CCN. Reuter and Guan (1995) gave the details of the parameterization scheme and justified its use for cloud pollution studies.

The partial differential equations are solved by a finite difference scheme on a staggered grid. A nondiffusive leapfrog method in time and centered difference method in space is used to solve the prognostic equations. The vertical grid spacing is uniform at 100 m. In the radial direction, the resolution is 83 m near the central axis, but widens smoothly towards the lateral boundaries. This grid stretching is computationally efficient, since coarser resolution is adequate near the lateral boundaries. A total of 125 horizontal mesh points are used to cover a radius of 20 km.

Sensible heat and water vapor release from industrial complexes were included using the approach of Murray *et al.* (1978), Orville *et al.* (1981), and Guan and Reuter (1995). The model convection was initiated by adding a small perturbation in the initial humidity field. Without such perturbation, there is no cloud formation in the sensitivity experiments for which the source of sensible and latent heat had been switched off, making comparisons among numerical experiments problematic. The humidity impulse was placed near cloud base height in the center of the domain and had a bell-shaped radial distribution. Its peak added about 15% in relative humidity compared with the environmental background value.

10.3 Factor Separation Methodology

We try to quantify the relative contributions of the sensible heat, latent heat, and industrial pollution on cumulus development and rainfall. The common technique used to assess the contribution of a specific factor is to compare the numerical results from the control run (which includes all factors) with the results from a model run in which the specific factor is "switched off". Thus, to determine the effects of sensible heat, two numerical experiments are made: the control run SLP, which including sensible heat (S), latent heat (L), and pollution (P); and the run LP, which includes latent heat and pollution, but not sensible heat. For any predicted variable f, the difference value $f(\text{SLP}) - f(\text{LP})$ is interpreted to quantify the effect of the sensible heat on f. It should be stressed here that if reference is made to the "latent heat" factor, *all* the effects of the additional water vapor are included.

Stein and Alpert (1993) criticized this common interpretation as $f(\text{SLP}) - f(\text{LP})$ includes not only the individual effect of the sensible heat, but also the *mutual effect* arising from the interaction of the sensible heat with the latent heat effect and the pollution effect. Instead they advocated the Alpert–Stein Factor Separation Methodology, which isolates the resulting values due to any factor in its pure form from the possible interactions with other factors. A total of eight numerical experiments are required: $f(\text{SLP})$, the predicted value of the control simulation including sensible heat, latent heat, and pollution; $f(\text{SL})$, the predicted value of the simulation when pollution is switched off; $f(\text{SP})$, the value of the simulation when latent heat is switched off; $f(\text{LP})$, the value of the simulation when sensible heat is switched off; $f(\text{S})$, the value when only sensible heat is switched on; $f(\text{L})$, the value when only latent heat is switched on; $f(\text{P})$, the value when only pollution is switched on; and $f(0)$, the value when all three factors are switched off (i.e., natural case unaffected by the industrial sources). The pure contributions by sensible heat, vapor, and pollution are $F_S = f(\text{S}) - f(0)$, $F_L = f(\text{L}) - f(0)$, and $F_P = f(\text{P}) - f(0)$, respectively. $F_{SL} = [f(\text{SL}) + f(0)] - [f(\text{S}) + f(\text{L})]$ is the contribution due to interaction of sensible heat and vapor. Similarly, $F_{SP} = [f(\text{SP}) + f(0)] - [f(\text{S}) + f(\text{P})]$, and $F_{LP} = [f(\text{LP}) + f(0)] - [f(\text{L}) + f(\text{P})]$ quantify the interactions between sensible heat and pollution, and vapor and pollution, respectively. The triple interaction of all three factors is given by $F_{SLP} = [f(\text{SLP}) + f(\text{S}) + f(\text{L}) + f(\text{P})] - [f(0) + f(\text{SL}) + f(\text{SP}) + f(\text{LP})]$. To compare the relative contributions of these terms it is convenient to express them in percentage terms; this is done by multiplying them by $100\%/f(0)$, where $f(0)$ is the value when all three factors are switched off.

It should be stressed that the results arising from the FS method depend to some extent on the number of factors being considered. Alpert *et al.* (1995) showed that adding a fourth factor in the FS analysis can erode the contribution associated with interactions of any two factors. However, when applying the technique to study the impact of the emissions of an oil refinery there are no obvious additional factors except for the three considered here. Therefore, the FS technique provides a useful approach for this study.

10.4 Model results

The relative importance of fluxes of sensible heat, latent heat, and pollution is determined for the case of the Wood River petroleum refinery. Auer (1976) estimated that this refinery emitted sensible heat and latent heat at rates of 0.4 gigawatt (GW) and 0.6 GW, respectively. However, for the purpose of quantifying the relative importance of sensible heat versus latent heat, it is convenient to have the same emission rate for the two forms of waste energy. That is, the rates of sensible heat and latent heat are assumed to be 0.5 GW. By having the same amount of energy, the difference between these two forms of energy becomes more apparent. The heat and vapor are released uniformly from a finite source region. Specifically the source region is a cylinder with base height 100 m, depth 100 m, and a diameter of 500 m. The cylinder is centered on the vertical z-axis and constitutes the "effective volume" of the heat and vapor plume interacting with the boundary layer. Both heat and vapor are distributed uniformly within the source cylinder. The pollution (i.e., CCN) effect enters the model via the specification of the cloud droplet concentration (N_c), the relative mass dispersion coefficient of the cloud droplet distribution (v_m), and the effective supersaturation value required for cloud condensation (S_p). The parameters were $N_c = 0.6$ per liter, $v_m = 0.7$ and $S_p = 0\%$ for polluted air, and $0.2\ l^{-1}$, 0.5, and 0.4% for unpolluted air, which were chosen to mimic the Wood River conditions as discussed by Reuter and Guan (1995). The atmospheric sounding used to initialize the cloud model was taken from Murray *et al.* (1978). The sounding had nearly calm conditions and supported shallow convection, which are prerequisites for using the assumption of axisymmetry.

Table 10.1 lists the eight numerical experiments made for each of the soundings. Each experiment is labelled with a code. The letters of the code indicate which of the three factors (sensible heat S, latent heat L, or pollution P) have been included in the simulation. Table 10.2 lists the domain-integrated mass of cloud water (CM) and rain water (RM) for the control experiment SLP and the "natural" case N (which has none of the three factors). The development of the "industrial" and the "natural" cumulus clouds is similar. The development of rain in N is slightly slower than that in SLP. Also the peak magnitudes of the integral variables are smaller in the "natural" cloud.

To evaluate quantitatively the effect of the industrial complexes on the convective rain shower, we compare the *maxima* of total cloud mass, rain mass, and accumulated rainfall at the surface (Table 10.3). The data listed can be interpreted as an intercomparison of the sensitivity experiments at the stage of maximum cloud development (i.e., maturing phase of the cumulus). Maximum values of SLP tend to be the largest, while those of N are the smallest. Moreover, maxima of SLP are much larger than those of N. Thus, the industrial complex intensifies the convective

Table 10.1 *List of numerical experiments. The input parameters are the emission rates of sensible heat (H$_S$) and latent heat (H$_L$), the cloud droplet concentration (N$_c$), the relative mass dispersion coefficient of cloud droplets (v$_m$), and the effective supersaturation (S$_p$).*

	SLP	N	SL	SP	LP	S	L	P
H_S (GW)	0.5	0.0	0.5	0.5	0.0	0.5	0.0	0.0
H_L (GW)	0.5	0.0	0.5	0.0	0.5	0.0	0.5	0.0
$N_c(l^{-1})$	0.6	0.2	0.2	0.6	0.6	0.2	0.2	0.6
v_m	0.7	0.5	0.5	0.7	0.7	0.5	0.5	0.7
$S_p(\%)$	0.0	0.4	0.4	0.0	0.0	0.4	0.4	0.0

Table 10.2 *Time evolution of cloud mass (CM) and rain mass (RM) in units of gigagram for the "industrial" case SLP and the "natural" case N*

Time	"Industrial" case (SLP)		"Natural" case (N)	
(min)	CM	RM	CM	RM
5	0.2	0.0	0.1	0.0
10	0.7	0.0	0.6	0.0
15	2.4	0.2	1.6	0.4
20	3.4	1.2	2.5	1.1
25	3.5	3.7	2.6	2.5
30	3.8	6.2	3.3	4.1
35	3.9	9.1	2.7	7.0
40	1.8	5.0	0.6	3.9
45	0.6	1.2	0.0	0.6
50	0.0	0.2	0.0	0.1

rain shower. The maxima of N are about the same as those of L, i.e., the latent heat emitted by the industrial complex has little effect on cloud development.

The results of the FS method for the maximum values of CM, RM, and AR are summarized in Table 10.4. Both sensible heat and pollution increase the maximum of cloud water mass. The contribution to RM$_{max}$ due to sensible heat (without interaction with other factors) is about 3.6 times larger than that of pollution. The difference between the effects of sensible heat and pollution on accumulated rainfall at the surface is even more obvious: the pure contribution of sensible heat increases AR$_{max}$ 7.5 times more than the pure pollution effect. The pollution has the largest effect in increasing CM$_{max}$. However, the sensible heat is dominant for enhancing

Table 10.3 *Comparison of maximum cloud water mass* *(CM*$_{max}$*), maximum rain water mass (RM*$_{max}$*), and* *maximum accumulated rain (AR*$_{max}$*) at the ground. All* *quantities are in units of Gg. Experiment N is "natural"* *case, which has no surface of waste heat, humidity, or* *pollution.*

	SLP	N	SL	SP	LP	S	L	P
CM_{max}	3.90	3.46	3.75	4.12	3.79	3.76	3.45	3.81
RM_{max}	9.10	6.90	8.86	8.91	7.33	8.48	6.89	7.34
AR_{max}	8.31	5.60	8.20	6.63	5.84	6.74	5.60	5.75

Table 10.4 *Contributions (in percentage relative to the case with all three factors* *switched off) to maximum cloud water mass (CM*$_{max}$*), rain water mass (RM*$_{max}$*)* *and accumulated rainfall (AR*$_{max}$*) by the pure effects F*$_S$*, F*$_L$*, F*$_P$*; interaction* *effects among two factors F*$_{SL}$*, F*$_{SP}$*, F*$_{LP}$*; the triple interaction effect F*$_{SPL}$*, and* *the total effect F*$_{TOT} = F_S + F_L + F_P + F_{SL} + F_{SP} + F_{LP} + F_{SPL}$*.*

	F_{TOT}	F_S	F_L	F_P	F_{SL}	F_{SP}	F_{LP}	F_{SLP}
CM_{max}	12.6%	8.5%	−0.4%	9.8%	0.1%	0.7%	0.0%	−6.1%
RM_{max}	31.8%	22.9%	−0.1%	6.4%	5.6%	−0.2%	−0.1%	−2.7%
AR_{max}	48.3%	20.3%	0.0%	2.7%	26.0%	−4.7%	1.7%	2.4%

the rainfall. This is consistent with the conclusion of Hindman *et al.* (1977b) that the pollution is not by itself responsible for the increased rainfall.

Comparing the pure contributions of the three factors, the pure effect of sensible heat is the most important for enhancing the rainfall. The dominant contribution of the sensible waste heat arises from the local enhancement of thermal buoyancy just beneath the cumulus cloud base. The heat flux above the source region yields a temperature perturbation of about 0.5 °C at cloud base, which supports a substantial cloud base updraft. The pollutant plays a significant role in increasing total cloud mass. The pure contribution of the latent heat effect is small on the convective rain shower. The contributions arising from the mutual interactions among two and three factors are not negligible. The contribution to CM_{max} from mutual interaction among sensible heat, latent heat, and pollution is comparable with the pure contribution of sensible heat or pollution, and is much larger than that of vapor. The contribution to RM_{max} due to the mutual interaction between sensible heat and latent heat is much larger than the pure effect of vapor. In fact, this mutual contribution to AR_{max} is the largest component (26%).

10.5 Discussion and conclusion

In order to better interpret the findings of the FS analysis, it is necessary to stress the major shortcomings of the present model approach. A key model assumption is that the rain formation in polluted air can be adequately modelled using a liquid-phase bulk-water parameterization scheme. The reason for excluding the ice phase in this study is the lack of understanding of how the ice-phase microphysics changes with extra anthropogenic aerosols arising from combustion of fossil fuels. It has been speculated that an increase of atmospheric sulfates stimulates rainfall through enhanced mixed-phase colloidial instability. The argument goes as follows. In a mixed-phase cloud, ice crystals grow by vapor diffusion at the expense of super-cooled water droplets. When the ice crystals have grown beyond about 200 μm they start to collect supercooled droplets on their way downward through the cloud. Clearly, the growth of these falling ice particles is affected by the cloud droplet distribution that depends on anthropogenic CCN such as sulfates. However, the presence of combustion pollution alters not only the concentration and distribution of CCN. However, little is known about how combustion pollution affects freezing nuclei. Furthermore, a change in the size distribution of primary ice crystals and supercooled cloud drops is likely associated with effects on ice multiplication processes. To avoid these complications, the present study has focused entirely on warm rain processes. The model soundings were selected to yield only shallow convection in which the model cloud top never reached the freezing level.

An axisymmetric cloud model is used to simulate a convective rain shower in calm wind conditions affected by an oil refinery that emits sensible heat, moisture, and industrial CCN. The FS method is used to quantify the pure and mutual interactive effects of sensible heat, vapor, and pollution. The cloud development is affected mostly by the release of sensible heat. The pure contribution of the sensible heat effect on accumulated rainfall at the surface and total rain water aloft is much larger than that of vapor or pollution. A tentative conclusion is that the sensible heat flux emitted from the refinery is the major contributing factor toward increased rainfall. However, pollution and sensible heat play an important role in increasing the amount of cloud water.

The contributions arising from mutual interactions among two or three factors are quite large and should not be neglected. Particularly, the interaction of the sensible heat and pollution effects contribute towards the accumulated rainfall.

11

The use of the Alpert–Stein Factor Separation Methodology for climate variable interaction studies in hydrological land surface models and crop yield models

D. Niyogi, R. Mera, Yongkang Xue, G. Wilkerson, and F. Booker

The Alpert–Stein Factor Separation Methodology (FS) method has been utilized in the study of the biophysical response to changes in the environment to assess the relative contribution of different atmospheric factors to the biological system. In this chapter we will discuss crop simulation and land surface model-based assessments of the sensitivity to past and future changes in climatic conditions: increasing CO_2, soil moisture, temperature and radiative conditions, and crop management procedures (irrigation). FS is applied to discern specific contributions to plant responses by single variables or combinations of environmental conditions. Our FS analysis has shown that it is important to understand that biological responses are inherently dependent on multiple variables in the natural world and should not be limited to assessments of single specific parameters.

11.1 Introduction

In this chapter we demonstrate how the FS analysis technique is a useful tool for crop–climate change (crop-clim) studies. Important interactions between the atmosphere and biophysical processes occur under land surface and atmospheric carbon dioxide (CO_2) level changes. We employ the Alpert–Stein FS Methodology (Stein and Alpert, 1993; Alpert, 1997) to investigate the direct as well as the interactive effects of soil moisture, temperature, and radiative changes on the direct effects of CO_2 doubling for different land-use/vegetation types, including agricultural production.

We present recent research using land surface and agrotechnology models that applied the FS technique to evaluate the direct and interactive effects of: (i) soil moisture changes on the biological effects of CO_2 doubling for different land-use/ vegetation types (Niyogi and Xue, 2006); (ii) temperature, radiation, and

Factor Separation in the Atmosphere: Applications and Future Prospects, ed. Pinhas Alpert and Tatiana Sholokhman. Published by Cambridge University Press. © Cambridge University Press 2011.

precipitation changes on two different crops (Mera *et al.*, 2006); and (iii) differing temperature, radiation, precipitation, and irrigation procedures on soybean production at ambient and enhanced CO_2 conditions (Mera *et al.*, 2010). This work used a photosynthesis-based gas-exchange/transpiration model (GEM, Niyogi *et al.*, 2009) coupled to an atmospheric-boundary layer model (Alapaty *et al.*, 1997), and CSM-CROPGRO-Soybean and CSM-CERES-Maize crop/agrotechnology models (Jones *et al.*, 2003; Hoogenboom *et al.*, 2004).

This chapter is arranged as follows. We briefly outline the models used in the various experiments in Section 11.2. Section 11.3 details the application of FS to diagnose the response to a doubling of CO_2 in a coupled atmosphere–biosphere model. In Section 11.4 we present how FS can be used to determine the relative contributions by climate variables and their interactions to maize and soybean crops and how these interactions and contributions may change in the presence of irrigation and enhanced CO_2 conditions. In Section 11.5 we present conclusions from these studies and propose methods by which FS can enhance the study of biosphere–atmosphere interactions to improve the techniques employed and model performance.

11.1.1 Background

Recent climate projections have continued to predict increasing atmospheric CO_2 and water vapor levels along with changes in surface temperature and rainfall patterns (IPCC, 2007). Curtis and Wang (1998) performed a comprehensive meta-analysis of over 500 reports of elevated CO_2 impacts on plant response and concluded that enhanced CO_2 levels led to a significant increase in total biomass and plant net carbon assimilation rates. As summarized in Beltrán-Przekurat *et al.* (this book, Chapter 6), Eastman *et al.* (2001) developed a regional-scale sensitivity study and concluded that the land-use land-cover change (LULCC) effects are comparable or even at times outweigh the direct CO_2 impacts. The biological impacts due to CO_2 changes are important, and could be even more dominant than the direct radiative effects of CO_2 changes within the surface meteorology/regional climate change framework (Pielke *et al.*, 2002).

Alteration in agricultural activity is a major driver of regional LULCC. Feedbacks that occur due to LULCC are noticed in shifts in regional climate patterns, which in turn lead to changes in the vegetation productivity and land surface feedback (Pielke *et al.*, 2002). As discussed in Foley *et al.* (2005), an increase in the human population and the demand for food and fiber has expanded the croplands, pasturelands, plantations, and urban areas in recent decades. Croplands and pastures presently occupy nearly half of the land surface and have become one of the largest land-use categories, rivaling forest cover in extent. Fall *et al.* (2010) concluded that the agricultural landscapes have contributed to regional cooling of the surface in recent decades. Global climate change is expected to increase agricultural yields in colder environments, but presents major challenges for crops grown in warmer climates or with limited water resources (Iglesias *et al.*, 1996; IPCC, 2007, Working Group II Report, p. 15; Hatfield, 2008). These assessments are based on field experiments that have analyzed effects of climate change on agricultural systems, usually using single factor approaches (Allen *et al.*, 1987, 2004; Lawlor and Mitchell, 1991; Jablonski *et al.*, 2002; Council for Agricultural Science and Techonology, 2004; Pritchard, 2005). Pielke *et al.* (2004) argue that the bidirectional effect of climate change on agriculture, and that of agricultural changes on regional climate need to be analyzed in future assessments. We show that the FS framework provides a good methodological basis for such studies.

11.2 Models

11.2.1 *Coupled biosphere–atmosphere model*

In the first study, we used a photosynthesis-based scheme coupled to land surface models. This approach is similar to the parameterizations and scaling discussed in SiB2 – Simple Biosphere Model, version 2 (Sellers *et al.*, 1996). The biophysical module was dynamically coupled with a prognostic soil moisture/soil temperature scheme (Noilhan and Planton, 1989), and an atmospheric boundary layer/meteorological model (Alapaty *et al.*, 1997). The transpiration/photosynthesis module is based on the Ball–Woodrow–Berry stomatal model (Ball *et al.*, 1987; Niyogi and Raman, 1997) and the Collatz *et al.* (1991, 1992) photosynthesis scheme. The stomatal conductance (g_s) (inverse of stomatal resistance, R_s) is estimated as

$$g_s = m \cdot \frac{A_n \cdot \text{rh}_s}{C_s} + b \tag{11.1}$$

where A_n is the net carbon assimilation (photosynthesis) rate, rh_s is the relative humidity, and C_s is the CO_2 concentration at the leaf surface. The terms m and b are constants based on gas-exchange considerations (Ball *et al.*, 1987) as a function of C3 or C4 vegetation and land use (Sellers *et al.*, 1996). The physiological variables such as A_n, C_s, and rh_s at the leaf surface are estimated using transpiration/ photosynthesis relationships at the leaf scale (Niyogi *et al.*, 2009).

Photosynthesis or carbon net assimilation is calculated as the difference between gross carbon assimilation (A_g) and respiration (R_d). Gross assimilation is taken as the minimum of three rates, limited by the photosynthetic (Rubisco) enzyme efficiency (Wc), the photosynthetically active radiation (PAR) captured by the leaf (We) and the leaf capacity to transport or adopt the photosynthetic outcome (Ws).

The atmospheric core and the surface models are similar to those described in Alapaty *et al.* (1997, 2001). Some changes were made to accommodate the photosynthesis and CO_2 effects within the program and the deep soil moisture and gravitational corrections to the force-restore method following Boone *et al.* (1999) to override the Jarvis-type canopy resistance.

11.2.2 CSM-CROPGRO-Soybean and CERES-Maize

We used the Decision Support System for Agrotechnology Transfer (DSSAT) models: CSM-CROPGRO-Soybean and CSM-CERES-Maize (Jones *et al.*, 2003; Hoogenboom *et al.*, 2004). The models simulate potential changes in the productivity and/or vegetation feedback as a function of environmental factors. The ability of CROPGRO-Soybean (SOYGRO in Hoogenboom *et al.*, 1992) and CERES-Maize (Hoogenboom *et al.*, 1994) to provide information on the impact of climate on crops is well documented. Indeed, the DSSAT models have been successfully applied to the study of the impact of climate change on various regions of the world (Wolf and van Diepen, 1995: Carlson and Bruce 1996; Hansen *et al.*, 1996,1999; Brown and Rosenberg, 1997; Lal *et al.*, 1998; Southworth *et al.*, 2000; Wang *et al.*, 2001; Magrín *et al.*, 2002; Wolf, 2002; Mall *et al.*, 2004; Mera *et al.*, 2006). A number of calibration and field validation experiments have aided such global application of the models (Boote *et al.*, 1997; Allen and Boote, 2000; Alagarswamy *et al.*, 2006). Following successful calibration, the analysis of climate change and related sensitivity studies for impact assessments are typically conducted. The model results are deemed realistic and representative of the environmental conditions studied. CSM-CROPGRO-Soybean is a predictive, deterministic model which simulates physical, chemical, and biological processes in the plant and its associated environment. The model simulates crop yields and related agronomic parameters and predicts primary plant processes based on weather, soil, and crop management conditions. The model is process-oriented and considers crop development, carbon balance, crop and soil nitrogen balances, and soil water balance (Boote *et al.*, 1998). Crop development in the model is sensitive to temperature, photoperiod, water deficit, and nutrient stresses during various growth phases and is expressed as the physiological days per calendar day (PD d^{-1}).

The CSM-CERES-Maize (Crop Environment Resource Synthesis-Maize; Jones and Kiniry, 1986; Jones *et al.*, 2003; Hoogenboom *et al.*, 2004) model is also a part of the DSSAT and is a predictive, deterministic model. The model is designed to simulate corn growth, soil, water, temperature, and soil nitrogen dynamics on a field scale for one growing season, and belongs to the same DSSAT family as CSM-CROPGRO-Soybean. CSM-CERES-Maize derives daily rates of crop growth (PGR, g plant^{-1} d^{-1}) as the product of light intercepted by the canopy

(IPAR, MJ plant^{-1} d^{-1}) and radiation use efficiency (RUE, g MJ^{-1}). When the crop is under environmental stress, this approach has limits in calculating photosynthesis and respiration (Lizaso *et al.*, 2005).

11.3 Application of factor separation for soil moisture and carbon dioxide effects on biological response

Niyogi and Xue (2006), utilized FS to extract the direct effect due to CO_2 change or soil moisture change alone and the interactive or indirect effects due to CO_2 and soil moisture change together. The hypothesis being tested was: the projected biophysical impacts of CO_2 changes are strongly dependent on the surface hydro-logical state as represented by the soil moisture conditions. Using the GEM model coupled to a PBL model, four combinations for two factors (soil moisture and CO_2) at two levels (soil moisture limiting and abundant) at present day and doubled CO_2 conditions were conducted. These experiments were performed for various land-use/land-cover conditions to extract the main effects (fCO$_2$ and fMoist) and their synergistic interaction (fMoist:CO$_2$). The FS equations were developed as:

$$
\begin{aligned}
f_0 &= F_0 \\
F_0 &\equiv (\mathrm{CO_2^-}, \mathrm{Moist^-})
\end{aligned}
\tag{11.1a}
$$

$$
\begin{aligned}
f_1 &= f_{\mathrm{CO_2}} = F_1 - F_0 \\
F_1 &\equiv (\mathrm{CO_2^+}, \mathrm{Moist^-})
\end{aligned}
\tag{11.1b}
$$

$$
\begin{aligned}
f_2 &= F_{\mathrm{Moist}} = F_2 - F_0 \\
F_2 &\equiv (\mathrm{CO_2^-}, \mathrm{Moist^+})
\end{aligned}
\tag{11.1c}
$$

$$
\begin{aligned}
f_{1,2} &= f_{\mathrm{Moist:CO_2}} = F_{1,2} - (F_1 + F_2) + F_0 \\
F_{1,2} &\equiv (\mathrm{CO_2^+}, \mathrm{Moist^+})
\end{aligned}
\tag{11.1d}
$$

In the above, CO_2^+ and CO_2^- refer to the scenario with doubling (68 Pa) and the then-present-day (34 Pa) climatological values of ambient CO_2 concentrations (e.g., Houghton *et al.*, 1996); and Moist$^+$, Moist$^-$, refer to the model setup for near-field capacity, or near-wilting soil moisture, respectively, for different land-use/vegetation types.

The results indicated that each of the land-use/vegetation types was unique in terms of its biological response and characteristics, yet some broad similarities could be identified. Accordingly, the different LULC categories could be clustered into four categories: Broadleaf Trees (SiB2 Vegetation Types 1, 2, and 3), Needleleaf Trees (SiB2 Vegetation Types 4 and 5), C4 Grass (SiB2 Vegetation Type 6), and C3 Grass and Shrubs (SiB2 Vegetation Types 7, 8, and 9). The results pertaining to these four groups of vegetation changes made within the model configuration are discussed in detail in Niyogi and Xue (2006). Figure 11.1 shows an example of the analysis for Vegetation Type 1. The figure portrays the average daytime variations in the simulated evapotranspiration (Etr) and net carbon assimilation or photosynthesis (An). Soil moisture dominated the Etr changes, while the CO_2 changes had a relatively small effect. Interestingly, the CO_2 impact on the Etr increased as soil moisture became limited. Changes in the photosynthesis rates were dominated by CO_2 levels (Fig. 11.1b), with higher CO_2 leading to higher An values under abundant soil moisture availability. When soil moisture was limiting, photosynthesis rates saturated around mid-day with the dip in the An curve generally corresponding to peak radiation values.

The model results showed different responses resulting from both CO_2 and soil moisture changes. This variability in the simulated outcome is due to both the direct effect as well as the interaction of CO_2 and soil moisture changes. Figure 11.2a, b gives the relative contribution of the direct changes in CO_2 and soil moisture, and their interactive feedback on modeled Etr and An.

In the figure, the first box-plot shows the direct biophysical effect of CO_2 changes, the middle corresponds to the direct effect of soil moisture changes, and the third corresponds to the interaction between soil moisture and CO_2. For Etr, some CO_2-based modulation was evident under limited soil moisture conditions. The corresponding FS results (Fig. 11.2a) suggest that the effect of doubling the CO_2 led to a reduction in Etr as a cumulative effect (interacting with soil moisture changes). Overall, the Etr changes were dominated by soil moisture availability, but the soil moisture–CO_2 interaction was antagonistic (i.e., opposite in sign as compared to the direct effects). The combined effects of soil moisture and CO_2 changes therefore suggest an overall reduction of the combined effect due to CO_2 rise and higher soil moisture availability. Considering the An response (Fig. 11.2b), the effect of CO_2 levels on carbon assimilation is clearly demonstrated (about ten times more effective than soil moisture). Again, for this case, the interaction term is antagonistic and suggests that, for the Broadleaf Evergreen trees, the impact of soil moisture availability on the net primary productivity or evapotranspiration reduces as CO_2 levels increase. As summarized in Niyogi and Xue (2006), these FS based results are physically realistic and consistent with different observations (e.g., Owensby et al., 1997; Curtis and Wang, 1998).

11.4 Factor separation for assessing climatic interactions on modeled soybean and maize yields

In Mera et al. (2006), the FS analysis was employed to understand the effect of individual as well as simultaneous changes in radiation (R), temperature (T), and precipitation (P) on simulated agricultural crop yields for maize and soybean.

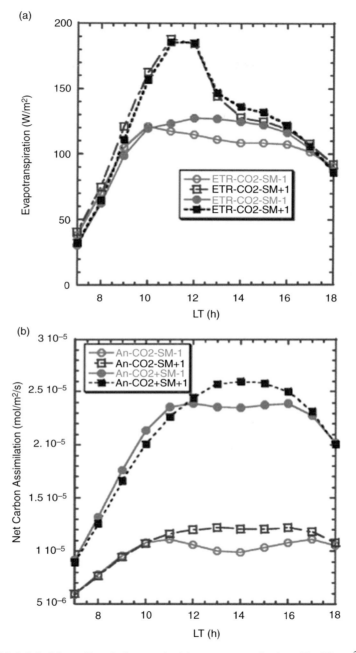

Figure 11.1 Model predicted changes in (a) evapotranspiration (Etr, W m^{-2}), and (b) net carbon assimilation (An, mol m^{-2} s^{-1}) for Vegetation Type 1 (Broadleaf-Evergreen trees). The high and low settings of the soil moisture and CO_2 values in the model initial conditions are represented by SM$^+$, SM$^-$, CO_2^+, and CO_2^-, respectively. Soil moisture has a dominant effect on Etr, while CO_2 changes are more important for An. From Niyogi and Xue (2006).

D. Niyogi, et al.

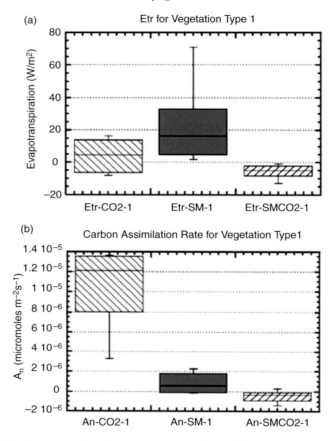

Figure 11.2 Box plots for factor-separated direct effects and interactions for (a) evapotranspiration (Etr, W m^{-2}), and (b) net carbon assimilation (An, mol m^{-2} s^{-1}) corresponding to Fig. 11.1. The SM:CO$_2$ term corresponds to the interaction effect. From Niyogi and Xue (2006).

Instead of eliminating or adding a factor as is generally done in the FS equations, for this study we used higher (p) and lower (m) settings of the climatic changes. Using the higher and lower values for the three variables, the eight FS equations were developed as:

$$E0 = mRmPmT \tag{11.2a}$$

$$ER = pRmPmT - mRmPmT \tag{11.2b}$$

$$EP = mRpPmT - mRmPmT \tag{11.2c}$$

$$ET = mRmPpT - mRmPmT \tag{11.2d}$$

$$\mathrm{E}RP = \mathrm{p}R\mathrm{p}P\mathrm{m}T - (\mathrm{p}R\mathrm{m}P\mathrm{m}T + \mathrm{m}R\mathrm{p}P\mathrm{m}T + \mathrm{m}R\mathrm{m}P\mathrm{m}T \qquad (11.2e)$$

$$\mathrm{E}RT = \mathrm{p}R\mathrm{m}P\mathrm{p}T - (\mathrm{p}R\mathrm{m}P\mathrm{m}T + \mathrm{m}R\mathrm{m}P\mathrm{p}T + \mathrm{m}R\mathrm{m}P\mathrm{m}T \qquad (11.2f)$$

$$\mathrm{E}PT = \mathrm{m}R\mathrm{p}P\mathrm{p}T - (\mathrm{m}R\mathrm{p}P\mathrm{m}T + \mathrm{m}R\mathrm{m}P\mathrm{p}T + \mathrm{m}R\mathrm{m}P\mathrm{m}T \qquad (11.2g)$$

$$
\begin{aligned}
\mathrm{E}RPT = {}& \mathrm{p}R\mathrm{p}P\mathrm{p}T - (\mathrm{p}R\mathrm{p}P\mathrm{m}T + \mathrm{p}R\mathrm{m}P\mathrm{p}T + \mathrm{m}R\mathrm{p}P\mathrm{p}T) \\
& + (\mathrm{p}R\mathrm{m}P\mathrm{m}T + \mathrm{m}R\mathrm{p}P\mathrm{m}T + \mathrm{m}R\mathrm{m}P\mathrm{p}T) - \mathrm{m}R\mathrm{m}P\mathrm{m}T) \qquad (11.2h)
\end{aligned}
$$

The terms on the left hand side of the equation are: E0, background effect or the model results, with the smaller prescribed values (m) of the R, P, T settings. Daily meteorological observations were modified as: $\pm 50\%$ of observed P, $\pm 25\%$ of observed R, and $\pm 2\,^{\circ}\mathrm{C}$ of observed T to get the smaller (m) and higher (p) settings. Parameter ranges were based on summary projections from climate model results and analysis of past regional climate data for seasonal variations (Mera *et al.*, 2006).

The terms such as ER, EP, and ET are the individual contributions or the direct effect of the variable R, P, and T, respectively. Terms such as ERP, ERT, and EPT are the double interactions between R and P, R and T, and P and T, respectively, while ERPT is the triple interaction effect due to combined changes in R, P, and T. The E in the equations represents the effect. The terms m and p represent the smaller ($-$) and the higher ($+$) values of the variable from the standard design of experiment perspective (e.g., Box *et al.*, 1978; Niyogi *et al.*, 1999).

In Fig. 11.3a, b, for example, the following information is included: (a) direct effect of individual variable changes, given as ER, ET, and EP; (b) the effect of interactions between two variables (e.g., temperature and radiation changing simultaneously), given as ERT, EPT, and ERP; and (c) the combined effect of all three variables simultaneously affecting the crop system, given as ERPT (cf. Eqs. 11.2 a–h). The results in the experiment showed that different combinations of changes in climate variable lead to significantly different responses. The impact of a variable change could also depend on the values of other variables, indicating a high degree of uncertainty in the crop yield projections under climate change conditions (Niyogi *et al.*, 1999). Using FS, the different characteristics of the interactions could be extracted and can help with understanding the effect and vulnerability associated with climate change and its potential impact on crop systems.

Figure 11.3a shows the factor separation plot for the CSM-CROPGRO-Soybean simulated soybean yield for a study domain configured and calibrated using chamber-grown soybean experimental data. The double interaction of radiation and

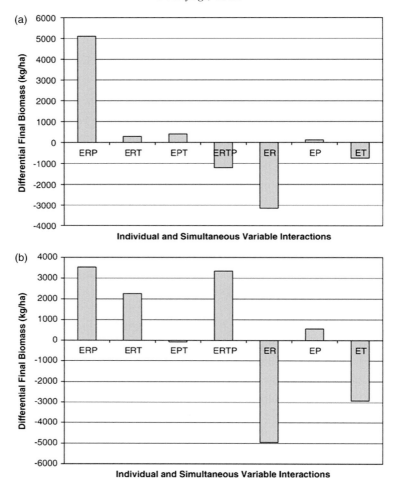

Figure 11.3 (a) Factor separation plot for soybean crop differential yield (kg ha^{-1}). A marked trend exists for the radiation–precipitation (EpRpT) interaction to have the largest positive effects. The radiation contribution (EpR) alone gives the lowest differential yield while the triple interaction (EpRpPpT) and temperature (EpT) also appear to have significant negative effects. (b) Same as (a), but for maize. From Mera *et al.* (2006).

precipitation (ERP) showed the largest positive effect, suggesting that increased radiation and precipitation would synergistically impact the yield in the model projections. Double interactions between temperature and radiation (ERT), and temperature and precipitation (EPT) have relatively smaller effects. The model captures the precipitation feedback as an interaction between precipitation and radiation. The triple interaction (ERPT) is also significant; however, this effect is smaller than the radiation–precipitation interaction. Thus, temperature changes can antagonistically interact with the dominant radiation–precipitation interaction. In the modeled yield estimates, radiation changes (ER) can cause a significant direct effect, indicating that some reduction in radiation values could aid crop growth, and this can be further enhanced by higher precipitation and lower temperature values.

The FS analysis of the simulated maize yield is shown in Fig. 11.3b. Radiation–precipitation interaction and radiation direct effects are the dominant factors within the model. These results are similar to those obtained in the soybean studies. That is, the radiation–precipitation interaction is strongly synergistic, and the radiation direct effect is a function of a negative feedback effect. Thus, up to a certain range, decreased radiation with increased precipitation provides the highest yields. The radiation feedback is nonlinear; i.e., relatively high and very low values could reduce the yield, and with average values the yield could be high. Two differences are seen for the soybean and maize simulation results. In maize, increasing temperatures show a positive interactive effect, as compared to the negative feedback in the soybean growth model. Additionally, as compared to soybeans, maize shows prominent interactions between radiation–temperature and temperature feedback. Therefore, the temperature feedback appears to be greater for simulating maize yields as compared to that of soybeans. Thus, unlike the soybean output, the maize output indicates that all R, P, T interactions make important contributions to maize growth. Such a scenario would lead to a more uncertain output in the model projections that is less vulnerable to individual changes, but more responsive to the system as a whole.

In a subsequent experiment currently being reported in Mera *et al.* (2010), the FS analysis was extended to examine the impact of simultaneous changes in weather variables and their interactions on soybean yield under ambient and enhanced CO_2 conditions as well as irrigated and nonirrigated fields. The data collected were also validated against container-grown soybean experiments by Booker *et al.* (2005). Similarly to Mera *et al.* (2006), the direct effects of variables and their interactions were calculated as:

$$E0 = (\mathrm{m}R\mathrm{m}P\mathrm{m}T) \tag{11.3a}$$

$$ER = (\mathrm{p}R\mathrm{p}P\mathrm{m}T) - E0 \tag{11.3b}$$

$$EP = (\mathrm{m}R\mathrm{p}P\mathrm{m}T) - E0 \tag{11.3c}$$

$$ET = (\mathrm{m}R\mathrm{m}P\mathrm{p}T) - E0 \tag{11.3d}$$

$$ERP = (\mathrm{p}R\mathrm{p}P\mathrm{m}T) - ER - EP - E0 \tag{11.3e}$$

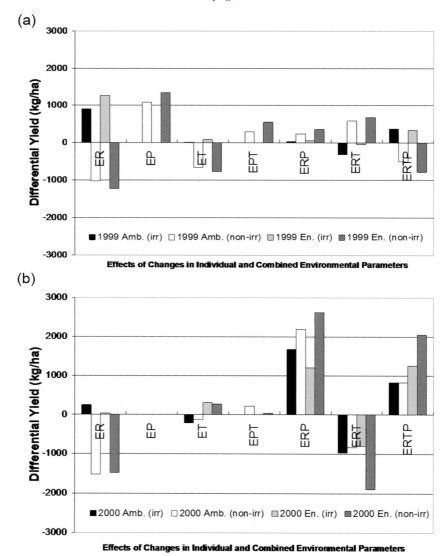

Figure 11.4 Factor separation plot for 1999 (a) and 2000 (b) soybean crop differential yield (kg ha^{-1}) Some of the main differences between the two seasons include the contributions by radiation–precipitation (ERP) and radiation–temperature (ERT) interactions. From Mera *et al.* (2010).

$$\mathrm{E}RT = (\mathrm{p}R\mathrm{m}P\mathrm{p}T) - \mathrm{E}R - \mathrm{E}T - \mathrm{E}0 \tag{11.3f}$$

$$\mathrm{E}PT = (\mathrm{m}R\mathrm{p}P\mathrm{p}T) - \mathrm{E}P - \mathrm{E}T - \mathrm{E}0 \tag{11.3g}$$

$$\mathrm{E}RPT = (\mathrm{p}R\mathrm{p}P\mathrm{p}T) - \mathrm{E}RP - \mathrm{E}RT - \mathrm{E}PT - \mathrm{E}R - \mathrm{E}P - \mathrm{E}T - \mathrm{E}0 \tag{11.3h}$$

The higher (p) and lower (m) settings were similar to those used in Mera *et al.* (2006). Figure 11.4 shows the effect of the weather variables and their interactions on yield.

The FS analysis showed that increases in radiation (ER) positively contributed to yield under irrigated conditions at both CO_2 levels. For nonirrigated (increased water stress) conditions, there was a negative effect on yield. If soil moisture levels were adequate, an increase in radiation for enhanced CO_2 conditions increased yield in the model compared with ambient CO_2 conditions. The FS results also highlighted that the CO_2 effects are sensitive to the availability of irrigation and could be potentially used for future climate change adaption/mitigation related studies.

The FS analysis allowed us to quantify the climate–crop interactions within the crop models and to further explore their role using scenario-based assessments.

11.5 Conclusions

The examples presented in this chapter provide insights into the use of the FS approach for studying climate–crop impact assessment studies. The scope of our studies ranged from the biological response to elevated levels of CO_2 for a variety of vegetation types, to isolating the important contributions from climate variables in growth and yield simulated by soybean and maize crop models. Application of the FS technique allowed us to extract the impact of simultaneous changes to environmental variables and their interactions as well as contributions made by single variable changes. The simulations in this study showed that changes in climate variables and interactions among these factors influence the extent of the transpiration, photosynthesis, and crop yield changes associated with increased CO_2. With growing emphasis on understanding the impacts of climatic changes on food security and ecosystem services, there would be increasing utility for using FS with land surface and crop models in the future.

Acknowledgements

This study was supported in part by the DOE ARM Program (08ER64674; Dr. Rick Petty), NSF CAREER (ATM-0847472, Dr. Liming Zhou and Dr. Jay Fein); NSF INTEROP (OCI 0753116, Dr. Sylvia J. Spengler), 480 the NASA-THP (NNG04G184G, Dr. Jared Entin), the NASA-IDS (NNG04GL61G, Drs. J. Entin and G. Gutman), USDA–NRICGP (through Tufts University, Dr. S. Islam), and Purdue University. The DSSAT data were made available through the ICASA.

12

Linear model for the sea breeze

T. Sholokhman and P. Alpert

Application of the Factor Separation (FS) technique is demonstrated with a simplified atmospheric problem with an analytical solution: the Haurwitz sea-breeze (SB) model. The synergy of chosen factors well explains a loop that the wind makes during a day at low latitudes with low friction. Only FS methodology gives the full picture of wind evolution in this case.

12.1 Modelling approach

The example of the FS methodology chosen is the dynamic model of the sea breeze suggested by B. Haurwitz (1947).

Haurwitz (1947) set up a simple SB model in which the spatial changes and changes in air compressibility were ignored. The equations of motion are as follows:

$$\frac{du}{dt} - fv + k_f u = P_x - F(t), \quad \frac{dv}{dt} + fu + k_f v = P_y \qquad (12.1)$$

where u and v are the x and y components of the surface horizontal wind, x is the horizontal axis positive from land to water, i.e., from west to east, y is the horizontal axis parallel to the shore from south to north, P_x and P_y are the components of large-scale pressure gradient force, $F(t)$ is a periodic function representing the pressure gradient between land and water that is caused by the diurnal variation of the temperature differences, $f = 2\Omega \sin\phi$ is the Coriolis parameter, and k_f is the friction coefficient. The general pressure gradient has the following components:

$$P_x = -\frac{1}{\rho}\frac{\partial p}{\partial x} = -fv_{gs}, \quad P_y = \frac{1}{\rho}\frac{\partial p}{\partial y} = fu_{gs} \qquad (12.2)$$

Factor Separation in the Atmosphere: Applications and Future Prospects, ed. Pinhas Alpert and Tatiana Sholokhman. Published by Cambridge University Press. © Cambridge University Press 2011.

where u_{gs} and v_{gs} are the x and y components of the geostrophic wind. $F(t)$ was chosen to be

$$F(t) = A/\pi + \frac{1}{2}A\cos\omega t \qquad (12.3)$$

where

$$A = \frac{1}{\rho}\frac{\partial p_0}{\partial x} = \frac{gz}{T}\frac{\partial T}{\partial x} \qquad (12.4)$$

A represents the relationship between horizontal temperature and pressure gradients, p_0 is the surface pressure.

Haurwitz (1947) shows that the solutions for u and v horizontal wind components are

$$
\begin{aligned}
u = &\frac{-k_f f}{f^2 + k_f^2}v_{gs} + \frac{f^2}{f^2 + k_f^2}u_{gs} - \frac{A}{\pi}\frac{k_f}{f^2 + k_f^2} \\
&- \frac{A}{2}\frac{(k_f^2 + \Omega^2 - f^2)\Omega\sin\Omega t + (k_f^2 + \Omega^2 + f^2)k_f\cos\Omega t}{\left(k_f^2 + \Omega^2 - f^2\right)^2 + 4k_f^2 f^2}
\end{aligned} \qquad (12.5)
$$

$$
\begin{aligned}
v = &\frac{f^2}{f^2 + k_f^2}v_{gs} + \frac{k_f f}{f^2 + k_f^2}u_{gs} + \frac{A}{\pi}\frac{f}{f^2 + k_f^2} \\
&- \frac{A}{2}\frac{(\Omega^2 - f^2 - k_f^2)f\cos\Omega t - 2fk\Omega\sin\Omega t}{(k_f^2 + \Omega^2 - f^2)^2 + 4f^2 k_f^2}
\end{aligned} \qquad (12.6)
$$

Haurwitz admits that the equations for u and v are difficult to discuss in general terms because of the auxiliary constants. He proceeds by taking a numerical example in which

$$A = 0.48 \text{ cm s}^{-1}$$

$$\Omega = 0.73 \times 10^{-4}\,\text{s}^{-1}$$

The Coriolis parameter of $f = 10^{-4}\,\text{s}^{-1}$ matches the latitude of $43°$. Also Haurwitz assumed that the coefficient of friction has the value of $k_f = 0.58 \times 10^{-4}\,\text{s}^{-1}$.

12.2 Factors chosen for the model

Suppose that the effects of the following two factors are investigated: the friction and the Coriolis force.

The Coriolis force is a fictitious force exerted on a body when it moves in a rotating reference frame. It is called a fictitious force since it is a by-product of measuring coordinates with respect to a rotating coordinate system as opposed to a real force. The Earth rotates about its axis from west to east once every 24 hours. Consequently, an object moving above the Earth in a generally northerly or southerly direction, and with a constant speed relative to space, will be deflected in relation to the rotation of the Earth. This deflection is clockwise, or to the right, in the Northern Hemisphere and anticlockwise, or to the left, in the Southern Hemisphere.

In physics, friction is the resistive force that occurs when two surfaces travel along each other when forced together. The surface of the Earth exerts a frictional drag on the air blowing just above it. This friction can change the wind direction and slow the near-surface air – keeping it from blowing as fast as the wind aloft. Actually, the difference in terrain conditions directly affects how much friction is exerted. For example, a calm ocean surface is pretty smooth, so the wind blowing over it does not move up, down, and around any features. By contrast, hills and forests force the wind to slow down and/or change direction much more. Hence, greater friction forces are exerted.

The two factors described above – the friction and the Coriolis force – were chosen for studying the FS Methodology. There are a number of studies of sea breeze that ignore both of these chosen factors (classical paper by Jeffreys, 1922). The FS method shows that these two factors have a great impact on the circulation of the sea breeze, and therefore cannot be disregarded.

The FS method allows examination of the changes in of the intensity and the direction of the wind close to the coast. These fluctuations are caused by the difference in latitude of the location, i.e., changes in the Coriolis term, and by different choices of the value of the friction term.

12.3 Factor separation results

Figure 12.1 shows typical hodographs of winds obtained by calculation when the geostrophic wind velocity is $0\,\mathrm{m\,s^{-1}}$.

The black diamond-shaped graph is the initial case with parameters:

$$k_f = 0.58 \times 10^{-4}\,\mathrm{s^{-1}}$$
$$f = 10^{-4}\,\mathrm{s^{-1}}$$

during the time from 0 to 24 h. The gray square-shaped graph is the second case where:

$$k_f = 10^{-4}\,\mathrm{s^{-1}}$$
$$f = 0.77 \times 10^{-4}\,\mathrm{s^{-1}}$$

where the Coriolis parameters match $43°$ and $32°$ latitudes correspondingly. The dots mark the end points of vectors with the origin at the intersection of the axes,

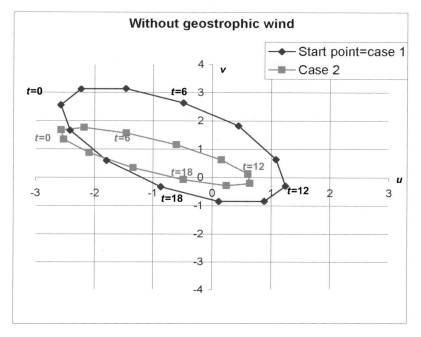

Figure 12.1 Hodographs of the theoretical sea breezes for zero geostrophic wind. The Black diamond-shaped graph is for parameters $k_f = 0.58 \times 10^{-4}\,\text{s}^{-1}$ and $f = 10^{-4}\,\text{s}^{-1}$, the gray square-shaped graph is for $k_f = 10^{-4}\,\text{s}^{-1}$ and $f = 0.77 \times 10^{-4}\,\text{s}^{-1}$. One side of a square corresponds to $1\,\text{m\,s}^{-1}$. In all cases land lies to the left and water to the right of the N-S line. The points indicate the (u, v) wind every 2 h starting at $t = 0$, which corresponds to maximum heating gradient (Eq. 12.3) or mid-day.

and the numbers represent the duration (in hours) from the moment of the maximum difference between the air temperatures over land and water.

In Fig. 12.1, which depicts sea breezes without the geostrophic wind, it can be clearly seen that on the Boston coast, latitude $43°$, the wind is generally stronger than at latitude $32°$. For example, at $t = 0$, just when the difference between the air and land temperature is maximal, the wind in Boston is $3.6\,\text{m\,s}^{-1}$, while in the second case it is $2.8\,\text{m\,s}^{-1}$. After 12 hours, the wind direction in Boston is south-west, while at latitude $32°$ the wind direction is north-west. Of course, it is not really a high-quality comparison, due to the different values of k_f.

The nonzero geostrophic wind greatly affects both the direction and the intensity of the wind. For example, consider the hodograph with the south geostrophic wind. In the general case, the same change is observed in both cases: the geostrophic wind turns the wind into its direction and intensifies the latter.

The south geostrophic wind produces the nonintersecting graphs (Fig. 12.2) where the Boston sea breeze is more intensive, and is directed more southward.

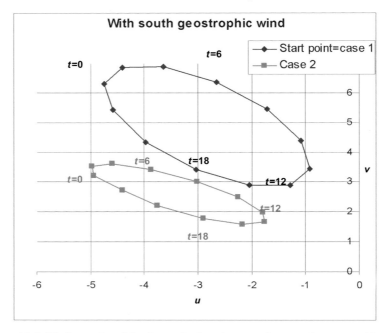

Figure 12.2 Hodographs of the theoretical sea breezes for a south geostrophic wind of speed $5 \, \text{m s}^{-1}$. The Black diamond-shaped graph is for parameters $k_f = 0.58 \times 10^{-4} \, \text{s}^{-1}$ and $f = 10^{-4} \, \text{s}^{-1}$, the gray square-shaped graph is for $k_f = 10^{-4} \, \text{s}^{-1}$ and $f = 0.77 \times 10^{-4} \, \text{s}^{-1}$. One side of a square corresponds to $1 \, \text{m s}^{-1}$. In all cases land lies to the left and water to the right of the N-S line. The points indicate the (u, v) wind every 2 h starting at $t = 0$, which corresponds to maximum heating gradient (Eq. 12.3) or mid-day.

We chose two factors for the FS method: the friction and the Coriolis force, while ignoring the nonzero geostrophic wind and time factor. For testing the synergism, four cases with different factors were chosen at time 0 and after 12 hours.

In Table 12.1, the vectors of the sea breeze represent the various choices of factors. It can be clearly seen that the winds are in accordance with the theory: they blow from sea to land – eastwards. After 12 hours from the time of maximum difference between the air temperatures over land and water, the winds blow from the land to sea (land breeze). Particularly striking is the weakening of the wind intensity in Table 12.2, which is apt to happen at night.

Figures 12.3 and 12.4 depict the wind vectors as a function of the Coriolis and the friction factors. The breeze becomes weaker and rotates clockwise as the Coriolis force increases from the southern latitude to the northern one. Obviously, the wind intensity increases with the decrease of friction. In this case, a clockwise motion occurs (also noticed in Tables 12.1 and 12.2):

Table 12.1 *Theoretical sea breeze for different values of factors k_f and f. The geostrophic wind is zero, time $t = 0$. The wind direction is reported with a meteorological angle; in all cases direction is south-east. Wind speed in m s^{-1}. In all cases land lies to the left and water to the right of the N-S line.*

f	k_f		
	0.3×10^{-4} s^{-1}	0.58×10^{-4} s^{-1}	10^{-4} s^{-1}
1.3×10^{-4} s^{-1}	160.1°	147.3°	135.4°
	3.5	3	2.4
10^{-4} s^{-1}	145.9°	134.8°	126°
	4.9	3.6	2.7
0.77×10^{-4} s^{-1}	122.4°	122.7°	117.9°
	5.6	3.9	2.9

Table 12.2 *Theoretical sea breeze for different values of factors k_f and f. The geostrophic wind is zero, time $t = 12$. The wind direction is reported with a meteorological angle; in all cases direction is west. Wind speed in m s^{-1}. In all cases land lies to the left and water to the right of the N-S line.*

f	k_f		
	0.3×10^{-4}s^{-1}	0.58×10^{-4}s^{-1}	10^{-4}s^{-1}
1.3×10^{-4}s^{-1}	328.6°	308.6°	294.6°
	1.3	1	0.6
10^{-4}s^{-1}	304.2°	283.1°	275.9°
	2.3	1.3	0.7
0.77×10^{-4}s^{-1}	263.6°	253.6°	258.3°
	3.4	1.4	0.6

12.4 A fractional approach to the Haurwitz SB solution

Here, we do not analyze the most basic zero case of the FS method, due to the fact that the theories of the sea breeze without the Coriolis force and friction were thoroughly discussed earlier, and are not very useful. The initial importance of the Haurwitz model was in understanding the important role of both the friction and the rotation of the Earth.

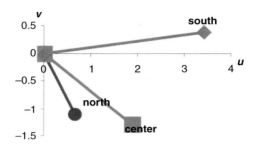

Figure 12.3 The breeze at $t = 12$ with friction coefficient 0.3×10^{-4} s^{-1}. The wind vector south has Coriolis parameter 0.77×10^{-4} s^{-1}, the central 10^{-4} s^{-1} and north 1.3×10^{-4} s^{-1}; these different Coriolis parameters correspond to latitudes of $32°$ (south), $43°$ (center) and $52°$ (north).

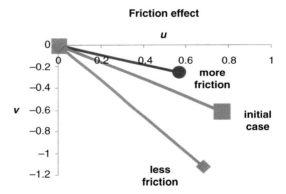

Figure 12.4 The breeze at $t = 12$ with Coriolis coefficient 1.3×10^{-4} s^{-1}. The wind vector with various friction coefficients is 10^{-4} s^{-1}, initial 0.58×10^{-4} s^{-1} and less friction 0.3×10^{-4} s^{-1}.

Our function with these two factors is

$$u(k_f, f) = \frac{-k_f f}{f^2 + k_f^2} v_{gs} + \frac{f^2}{f^2 + k_f^2} u_{gs} - \frac{A}{\pi} \frac{k_f}{f^2 + k_f^2}$$
$$- \frac{A}{2} \frac{(k_f^2 + \Omega^2 - f^2)\Omega \sin \Omega t + (k_f^2 + \Omega^2 + f^2) k_f \cos \Omega t}{\left(k_f^2 + \Omega^2 - f^2\right)^2 + 4 k_f^2 f^2} \tag{12.7}$$

$$v(k_f, f) = \frac{f^2}{f^2 + k_f^2} v_{gs} + \frac{k_f f}{f^2 + k_f^2} u_{gs} + \frac{A}{\pi} \frac{f}{f^2 + k_f^2}$$

$$-\frac{A}{2} \frac{(\Omega^2 - f^2 - k_f^2) f \cos \Omega t - 2 f k \Omega \sin \Omega t}{(k_f^2 + \Omega^2 - f^2)^2 + 4 f^2 k_f^2} \qquad (12.8)$$

where the basic-state (zero) position is assumed to be with:

$$k_f = 0.58 \times 10^{-4} \, \text{s}^{-1}$$
$$f = 10^{-4} \, \text{s}^{-1}$$

and we examine the cases with different friction coefficients while the Coriolis coefficient varies accordingly with transition from the Equator to the North Pole. We choose four cases for friction with increasing k_f:

1. $k_f = 0.5 \times 10^{-5} \, \text{s}^{-1}$

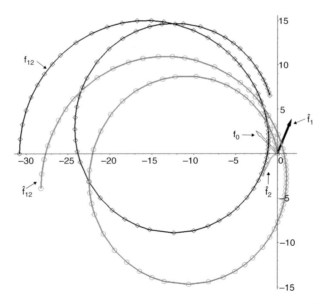

Figure 12.5 The breeze at $t = 0$ with friction coefficient $k_f = 0.5 \times 10^{-5} \, \text{s}^{-1}$. These different Coriolis parameters correspond to latitudes from $0°$ to $90°$. The diamond-shaped graph is the trajectory of final vector sea breeze in final choices f_{12}, the circle-shape is the trajectory of the synergic vector $\hat{f}_{12}$, the smooth-boxed-shaped line is the trajectory of the synergic effect of the Coriolis force $\hat{f}_2$, the dark-shaded vector is the synergic effect of the friction force $\hat{f}_1$ and the unshaded vector is the initial position f_0.

2. $k_f = 0.5 \times 10^{-4}\,\mathrm{s}^{-1}$

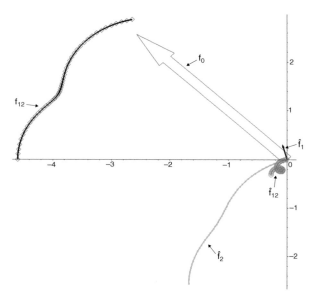

Figure 12.6 The breeze at $t = 0$ with friction coefficient $k_f = 0.5 \times 10^{-4}\,\mathrm{s}^{-1}$. Explanation as for Fig. 12.5.

3. $k_f = 10^{-4}\,\mathrm{s}^{-1}$

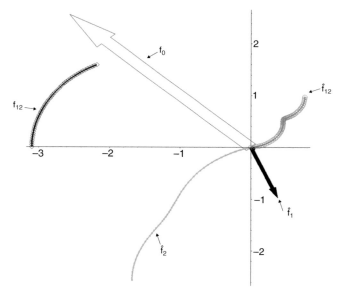

Figure 12.7 The breeze at $t = 0$ with friction coefficient $k_f = 10^{-4}\,\mathrm{s}^{-1}$. Explanation as for Fig. 12.5.

4. $k_{\mathrm{f}} = 0.5 \times 10^{-3}\,\mathrm{s}^{-1}$

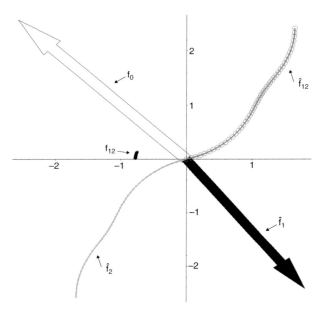

Figure 12.8 The breeze at $t = 0$ with friction coefficient $k_f = 0.5 \times 10^{-3}\,\mathrm{s}^{-1}$. Explanation as for Fig. 12.5.

12.5 The low-latitude phenomena

During analysis of the Haurwitz model, an anomaly was found. At low latitudes with low friction, during the day the wind makes a loop between $7°$ and $27°$, as distinct from the regular behavior of the breeze at night when it is calculated with large friction. From first analysis, it is seen that the loop of the wind vectors is formed by synergy of the friction and Coriolis forces, and not from the pure effects of each of these factors acting alone.

12.6 Summary

A strong feature of FS method is that it allows the investigation of nonlinear synergic terms directly. In this chapter, we investigate the synergy with a simple sea breeze and the theoretical early study by Haurwitz (1947). The synergic contribution due to the Coriolis/friction terms is found to be important, and its characteristics are exemplified.

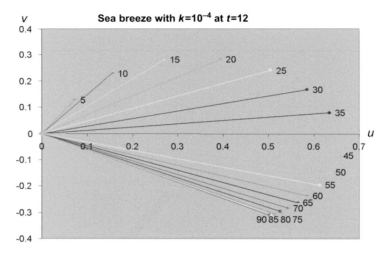

Figure 12.9 The breeze at $t = 12$ with friction coefficient $k_f = 10^{-4}\,\text{s}^{-1}$. The wind vector applied to different Coriolis parameters correspond to latitudes of $0°$ to $90°$. See plate section for color version.

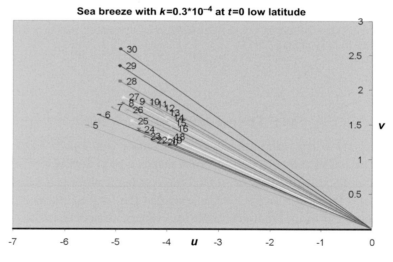

Figure 12.10 The breeze at $t = 0$ with a friction coefficient $0.3 \times 10^{-4}\,\text{s}^{-1}$. The wind vector applied to different Coriolis parameters corresponds to the latitudes of $0°$ to $30°$. See plate section for color version.

The SB wind behavior that is predicted by the model is in accordance with general observations. One of these observations is that in the Northern Hemisphere the wind tends to rotate clockwise.

During the day, looking from south to north, when the Coriolis force increases, the eastern sea breeze deflects more and more to the right side and becomes more southern. During the night, if we move from south to north, the western land breeze deflects to the right side and becomes more southwards.

The second common sense rule is trivial: friction slows the wind down. An important finding of our research is that, while the wind intensity at day is stronger than at night, the effect of pure friction is more dominant, especially at night. Also, when day and night are compared, the dominance of the synergy between the two factors becomes substantial at night.

Concluding these findings, the two chosen factors and their chosen fractional changes have a lesser influence on the sea breeze during the day, while at night the influence is significant.

The synergy of these chosen factors fully explains the loop that the wind makes during a day at low latitudes with low friction. Only FS methodology gives the full picture of the wind evolution in this case.

13

Experience and conclusions from the Alpert–Stein Factor Separation Methodology

Ensemble data assimilation and forecasting applications

D. Rostkier-Edelstein and J. P. Hacker

A long-term goal of our work is to find an efficient system for probabilistic planetary boundary layer (PBL) very short-range forecasting (nowcasting) that can be employed wherever surface observations are present. Our approach makes use of a single column model (SCM) to predict the state of a PBL column, and an ensemble filter (EF) to assimilate surface observations. Although the system may be run under different levels of complexity it is not immediately clear that the additional complexity will improve the performance of a PBL ensemble system based on a simple model. We address this question in the frame of Alpert–Stein Factor Separation Methodology (ASFSM) analysis with regard to treatment of parameterized radiation, horizontal advection and assimilation of surface observations. Results show that the added complexity often improves the forecasts under most skill metrics, but that assimilation of surface observations is the most important contributor to improved skill.

13.1 Introduction

Successful nowcasting and forecasting of the state of the PBL is of value for a wide range of practical forecasting applications, and the aim of this work is to find an efficient method for probabilistic nowcasting (0–3 h forecasting) of the state of the PBL wherever surface observations are available. Convective initiation and forecasted precipitation have shown sensitivity to the PBL structure (Crook 1996; McCaul and Cohen, 2002; Martin and Xue, 2006). Air-quality analysis and plume dispersion can benefit from accurate PBL analysis of stability and mixing depth (e.g., Kumar and Russel, 1996; Shafran *et al.*, 2000). Wind resource siting and real-time wind power operations can take advantage of accurate wind analysis and forecasts at several heights within the PBL (Wagner *et al.*, 2008). Accurate

Factor Separation in the Atmosphere: Applications and Future Prospects, ed. Pinhas Alpert and Tatiana Sholokhman. Published by Cambridge University Press. © Cambridge University Press 2011.

representation of the initial PBL structure could lead to improved short-range model forecasts of local thermally driven circulations such as sea–land and mountain breezes (e.g., Leidner *et al.*, 2001).

Surface-layer *in-situ* observations (2 m shelter and 10 m anemometer height) are a dense, inexpensive, accurate and reliable data source. However, as discussed by Hacker and Snyder (2005, hereafter referred to as HS05) and Hacker and Rostkier-Edelstein (2007, hereafter referred to as HR07), several difficulties limit their optimal utilization in present operational numerical weather prediction (NWP) and data assimilation (DA) systems. Transient coupling and decoupling of the near-surface atmosphere with the surface and the atmosphere aloft lead to intermittent, anisotropic and nonstationary correlations of the observations with the model column above it, making it difficult to quantify the optimal vertical influence of the surface observations. Error growth of the background state is known imperfectly, is highly variable, and is usually not well represented in current mesoscale models. The inability to correctly estimate the representativeness error of the surface observations and the dynamic balances often exploited in large-scale DA systems (e.g., three- and four-dimensional variational approaches, 3DVar and 4DVar), which are inappropriate for surface observations, largely prevent their optimal assimilation.

HS05 and HR07 proved that EFs offer a method for overcoming many of the limitations suffered by assimilation of surface observations because they naturally use anisotropic and flow-dependent covariance estimates derived from an ensemble. They avoid the need to prescribe the vertical influence of the surface observations, as well as the use of large-scale dynamic balances. In addition, the ensemble provides a flow-dependent estimate of the uncertainty in the analyses and forecasts, thus offering a probabilistic analysis and forecast.

HR07 demonstrated that surface observations can be a significant source of information toward determining the state of PBL profiles. They assimilated surface observations into a simple single column model (SCM) of the PBL using an EF. Since the main interest is in vertical structures and relationships within and near the PBL, an SCM allows experimentation at a fraction of the cost associated with a 3D mesoscale model. Comparisons against free-running simulations (i.e., without the assimilation of surface observations), representing a "climatological" distribution, showed that, without additional sources of information, it is possible to estimate the lowest few hundred meters in the atmosphere with an accuracy that approaches observation-error levels.

Their work has been extended to quantify the deterministic and probabilistic skill of the same SCM, and the SCM with added complexity (Hacker and Rostkier-Edelstein, 2008; Rostkier-Edelstein and Hacker, 2009; and Rostkier-Edelstein and Hacker, 2010, hereafter referred to as HR08, RH09 and RH10, respectively). While it seems attractive to add additional physics and dynamics to the SCM model, expecting that these will make solutions more realistic and reduce errors that arise from the imperfect model, it is not immediately clear that additional complexity will improve the performance of a PBL nowcasting system based on a simple model and ensemble DA. This question is investigated with regard to two model components and their importance relative to surface assimilation. The Alpert–Stein Factor Separation Methodology (ASFSM) is useful in this case as it quantifies the individual contribution of each model component to the probabilistic performance of the system, as well as any beneficial or detrimental interactions between them.

The ASFSM has been applied to several sensitivity studies (Sholokhman and Alpert, 2007, some of them related to our study (Levy and Ek, 2001; Lynn *et al.*, 2001; Mera *et al.*, 2006). A study closely related to ours is that by Levy and Ek (2001), who used the ASFSM to investigate the response of the PBL to surface fluxes, prescribed advection and radiative forcing in a SCM initialized with observed sounding profiles. The present work is the first to make use of the ASFSM in the framework of deterministic and probabilistic verification of EF assimilation analyses and forecasts. It is also the first to consider statistical factors (prediction skill) as opposed to a physical response.

13.2 Modeling approach

13.2.1 The single column model

An SCM simulates or predicts the atmosphere, earth surface and soil in a single grid column and is forced by observations or gridded fields (e.g., mesoscale analyses or forecasts). SCMs have proven useful for specific nowcasting applications (see, e.g., Müller *et al.*, 2007; Terradellas and Cano, 2007; Rémy and Bergot, 2010[1]; and references therein).

Details about the SCM implementation are described in detail in HR07 and in RH10. It contains the same PBL, land surface, and soil parameterization schemes as the Advanced Research WRF (ARW) version 2.2 modeling system (Skamarock *et al.*, 2005). The SCM dynamics are momentum, thermodynamic, and moisture equations:

$$\frac{\partial U}{\partial t} = f(V - V_g) - \frac{\partial}{\partial z}(u'w') + \frac{U_u - U}{\tau_a} \tag{13.1}$$

$$\frac{\partial V}{\partial t} = f(U - U_g) - \frac{\partial}{\partial z}\langle v'w' \rangle + \frac{V_u - V}{\tau_a} \tag{13.2}$$

---

[1] This work makes use of an SCM and EF DA of various types of observations to nowcast fog. The effect of several system components is considered but without following the factor separation method.

$$\frac{\partial \Phi}{\partial t} = \frac{\partial}{\partial z} \langle w'\theta' \rangle + \frac{1}{\rho c_p} \frac{\partial R}{\partial z} + \frac{\Theta_u - \Theta}{\tau_a} \qquad (13.3)$$

$$\frac{\partial Q_v}{\partial t} = -\frac{\partial}{\partial z} \langle w'q_v' \rangle + \frac{Q_{vu} - Q_v}{\tau_a} \qquad (13.4)$$

Variables $\psi = (u, v, \theta, q_v)$ have been decomposed into mean (Ψ) and turbulent (ψ') components so that $\psi = \Psi + \psi'$ and the usual Reynolds averaging has been applied. Angle brackets denote an average over subgrid-scale eddies. The turbulent components are derived from PBL parameterizations. Tendencies from radiation schemes (RRTM long-wave (Mlawer *et al.*, 1997) and Dudhia short-wave (Dudhia, 1989) formulations) appear on the right-hand side of Eq. (13.3) as a function of radiative flux, or the vertical gradient of absorbed irradiance (R) from all wavelengths considered. The radiation schemes are incorporated to improve night-time radiative cooling. When these are not being used, radiative forcing is specified from 3D WRF model forecasts.

The right-most terms in Eqs. (13.1)–(13.4) represent relaxation terms that simulate tendencies due to horizontal advection. These are formulated based on an approach by Ghan *et al.* (1999). Variables ψ_u are upstream values determined by a 3D model (WRF in our specific implementation), and $\tau_a = \Delta X / |V|$ is the advective time scale given by the wind speed and the grid spacing of the 3D model (here $\Delta X = 4$ km). Thus, it relaxes the SCM state toward a prescribed (time-dependent) 3D state on the advective time scale.

Large-scale wind forcing is accounted for by prescribing geostrophic wind components (U_g, V_g) from the same 3D WRF simulations. These will dominate in weak turbulence and weak advection scenarios. Resolved dynamics are integrated in time with a Crank–Nicholson algorithm.

13.2.2 The use of ensemble filter data assimilation (EF DA)

Atmospheric EF DA and DA in general have been widely reviewed (see, e.g., Daley, 1991; Talagrand, 1997; Kalnay, 2003), and we will briefly summarize the essence of these. DA in geophysical fluid dynamics consists of the process by which all the available information and its uncertainty is used to estimate the present state of the atmospheric or oceanic flow (referred to as analysis in a deterministic sense, or posterior in a Bayesian probabilistic sense) as accurately as possible. The available information consists of present information provided by observations, and past information provided by model dynamical calculations and past observations (the past information is referred to as first guess, background, or prior). In matrix-vector notation the analysis can be expressed as

$$\mathbf{x}^a = \mathbf{x}^b + \mathbf{K}(\mathbf{y}^o - \mathbf{H}\mathbf{x}^b) \qquad (13.5)$$

The vectors $\mathbf{x}^a$, $\mathbf{x}^b$, and $\mathbf{y}^o$ are the updated state, the background (prior) state and the observations, respectively. $\mathbf{H}$ is a linear forward operator that transforms from model variables to observation space. Here $\mathbf{H}$ is the application of surface-layer similarity in the PBL parameterizations, giving surface meteorology predicted by the model. $\mathbf{K}$ is a weights matrix that adds the vector of "innovations" $(\mathbf{y}^o - \mathbf{H}\mathbf{x}^b)$ to the background state to get the updated state. Equation (13.5) is called the statistical analysis equation, and all sequential data assimilation systems can be written this way as long as the errors statistics are normal and linear, and the forward operator is linear.

$\mathbf{K}$ determines the weighting between the background and observations, and is a function of the estimated error covariances of the forecast and the observations:

$$\mathbf{K} = \mathbf{P}^b \mathbf{H}^T (\mathbf{H}\mathbf{P}^b\mathbf{H}^T + \mathbf{R})^{-1} \qquad (13.6)$$

where $\mathbf{P}^b$ and $\mathbf{R}$ are the background and observation error covariance matrices respectively.

In EF DA a short-range ensemble forecast is used to estimate the flow-dependent background error covariance for updating each of the ensemble members; it is a Monte-Carlo technique. HS05 discussed in detail the pros and cons of the various DA methods when assimilating surface observations. Although any data-assimilation system can use flow-dependent covariances, most implementations of 3DVar and 4DVar use stationary and isotropic $\mathbf{P}^b$, which can be problematic in the diurnally varying PBL. Also, the cost of the DA system rises rapidly when one combines the minimization in 3DVar or 4DVar with the cost of an ensemble forecast required to estimate a flow-dependent $\mathbf{P}^b$. Currently in most numerical weather prediction applications, the cost of 4DVar and an EF are similar, but in the EF we gain the ensemble prediction system as well. EF DA offers additional benefits (some of them enumerated in Section 13.1): it is technically easier to implement as it does not require the development of tangent-linear and adjoint models; it does not require the linearization of the evolution of the forecast error covariance (as ensemble 4DVar does); and it does not rely on large-scale balances (constraints applied in many Var systems). EF DA is not free of sources of error, some of which are discussed in, e.g., Anderson (2007). Some of these limitations are handled in our specific implementation (Section 13.3).

13.2.3 Probabilistic forecasting

The introduction of the EF DA method carries with it the concept of ensemble and probabilistic forecasts. A comprehensive discussion on ensemble forecasting can be found, e.g., in Kalnay (2003). In our context it is important to mention that ensemble forecasting provides a practical method to estimate the inherent flow-dependent uncertainty that accompanies any analysis or forecast. Forecast error growth is fast in the relevant PBL scales, and an estimate of the uncertainty is important even for very short-term forecasts. Moreover, ensemble-mean forecasts extend the range of skilful prediction as the most uncertain components tend to average out.

In a similar sense that verification is required to assess the value of a deterministic forecast, no estimate of the uncertainty in the forecast is trustworthy unless it is verified. Probabilistic techniques have been developed to verify the truthfulness of the probability distribution of states generated by the ensemble systems (see, e.g., Jolliffe and Stephenson, 2003). In our work we analyze both the deterministic and probabilistic performance of our system (Sections 13.5 and 13.6).

13.3 Specific SCM and EF implementation

The specific SCM and EF implementations are summarized in detail in HR07 and RH10 and we present here the most relevant information.

The SCM is coupled to the Data Assimilation Research Testbed (DART), a comprehensive EF system designed for research and education on ensemble filters with a variety of dynamical systems, and developed at the National Center for Atmospheric Research (NCAR). Many filter algorithms are available in DART and we chose the default ensemble adjustment Kalman filter algorithm as documented in Anderson (2001, 2003).

The coupled SCM-EF system can be applied at any location at which surface observations are available. In our present work we locate the SCM over a column near Lamont, Oklahoma, at the Southern Great Plains (SGP) Central Facility of the Atmospheric Radiation Measurement (ARM) program. The SGP ARM site is attractive because it is well instrumented, providing observations for both assimilation and verification of nowcast profiles. Three-dimensional model data over this area needed for initialization and forcing are available from real-time runs produced to support the Bow Echo and Mesoscale Convective Vortex Experiment (BAMEX) period spanning 3 May through 14 July 2003. WRF 36 h forecasts were launched at 0000 UTC every day, on a $\Delta X = 4$ km grid.

The ensemble-mean initial conditions, large-scale forcing, advective tendencies and surface radiation are specified equal to a WRF forecast valid for a given day and hour (corresponding to the valid time of the surface observations). Perturbations are created by adding to the mean the scaled difference between that forecast and another forecast from the archive. This additional member is selected randomly from the experiment period, and the scaling of the difference is drawn randomly from a normal distribution $N(0,1)$. This approach enables the creation of arbitrarily large ensembles, and we use 100 members in our experiments. The same weights and WRF forecasts are used for both the initial conditions and the forcing, so that forcing time series and initial conditions are consistent.

Ensembles are derived from a distribution of WRF forecasts with wider variance than the real analysis and forecast error, potentially introducing too much spread in the ensembles relative to ensemble-mean error. Although the assimilation of surface observations will contract the distribution quickly (cf. HR07), the advective tendencies can lead to unrealistically rapid growth in ensemble spread. In an effort to avoid this we adopt a state-augmentation approach: advective tendencies are included in the state vectors ($\mathbf{x}^b$ and $\mathbf{x}^a$ in Eq. (13.5)). Thus, the advection speed is dynamically tuned with the surface observations simulating the effect of assimilating data in three spatial dimensions. Additional modeling details, such as the vertical covariance localization procedure needed in EF DA to mitigate sampling error, are described in HR07 and RH10.

13.4 Experiment design and factor separation analysis

HR07 found that EF assimilation of surface observations into a simple SCM initialized and forced with climatological data yields quite accurate PBL profiles. Ensemble-mean analyses were verified against surface and independent rawinsonde observations. The system performance showed a diurnal and vertical height variation that depended on the analyzed variable. That SCM implementation lacked two of the model components described in Section 13.2.1: parameterized radiation and advective tendencies. HR08, RH09 and RH10 investigated the added value of these two components relative to surface assimilation when the SCM is initialized and forced with the most recent WRF forecast. Their study investigates the effect of these model components on the deterministic and probabilistic skill of the system through verification of the ensemble mean and of the distribution of ensemble members. The rest of this chapter discusses their work and results in the context of the ASFSM.

Implementation of the ASFSM requires examination of the system performance for all 2^3 possible model configurations associated with the three studied components of assimilation, radiation, and advection. The possible combinations are listed in Table 13.1 for reference.

Table 13.1 *Key to all possible combinations of the model components considered here*

Label	Advection	Radiation	Assimilation
000	No	No	No
100	Yes	No	No
010	No	Yes	No
001	No	No	Yes
110	Yes	Yes	No
101	Yes	No	Yes
011	No	Yes	Yes
111	Yes	Yes	Yes

Factor separation equations are derived by a Taylor expansion of the effect of each component. Stein and Alpert (1993) derived the resulting equations for three components (Eqs. (17)–(24) in their manuscript). In the context of our study, the variables e_{ijk} (with ijk the possible combinations in Table 13.1) are performance, or skill, of each of the 2^3 possible model configurations as measured by objective verification metrics (Section 13.5). The factors f_{ijk} affect the system's performance resulting from single, dual, and triple synergistic interactions among the three studied model components. Rearranging those equations allows us to quantify the effect of adding new system components or sets of components. We then define the variables ef_{ijk} as the "effect" of a given factor, f_{ijk}, on the performance (skill) of base system configuration (i.e., of e_{000}, the performance of the system without the inclusion of any of the studied components) as follows:

$$ef_{000} = f_{000} = e_{000} \tag{13.7}$$

$$ef_{100} = f_{000} + f_{100} = e_{100} \tag{13.8}$$

$$ef_{010} = f_{000} + f_{010} = e_{010} \tag{13.9}$$

$$ef_{001} = f_{000} + f_{001} = e_{001} \tag{13.10}$$

$$ef_{110} = f_{000} + f_{110} = e_{110} - (e_{100} + e_{010}) + 2 \cdot e_{000} \tag{13.11}$$

$$ef_{101} = f_{000} + f_{101} = e_{101} - (e_{100} + e_{001}) + 2 \cdot e_{000} \tag{13.12}$$

$$ef_{011} = f_{000} + f_{011} = e_{011} - (e_{010} - e_{001}) + 2 \cdot e_{000} \tag{13.13}$$

$$ef_{111} = f_{000} + f_{111} = e_{111} - (e_{110} + e_{101} + e_{011}) + (e_{100} + e_{010} + e_{001}) \tag{13.14}$$

Equation (13.7) represents the joint effect of all model components that are not analyzed in the present study on the system's performance (and it also represents the system's skill itself). Equations (13.8) and (13.9) show the pure effect of each single model component evaluated in this study. In these cases the values of ef_{ijk} represent also the measured performance of the system (e_{ijk}) when each individual component is included in the system, and in the absence of second or third components.

In contrast, Eqs. (13.11)–(13.14) show the effect of synergistic factors on the system's performance. These values of ef_{ijk} do not correspond to the measured performance of any configuration and they cannot be directly computed from the output of individual simulations. Rather, they show the change in performance of the base configuration due to nonlinear interactions between the studied model components. These dual and triple interactions are obtained when the system is run in a configuration that includes more than one of the studied components. The presentation of the ASFSM in terms of the variables ef_{ijk} is also convenient to precisely assess the statistical significance of the several effects (relative to the base configuration) in a straightforward manner through the calculation of confidence intervals (CIs) following a bootstrap technique (Section 13.5).

Experiments are designed to evaluate the deterministic and probabilistic skill of day and night-time soil and atmospheric conditions. SCM simulations are initialized at 0300 and 1500 UTC. Assimilation of surface observations is accomplished every 30 minutes, starting at 0400 and 1600 UTC, respectively (in any model configuration that includes assimilation). Observations for assimilation are 30-minute averages of temperature and water vapor mixing ratio at $z = 2$ m above ground level i.e., (AFL) (T_2, Q_2), and of winds at $z = 10$ m AFL, i.e., (U_{10}, V_{10}). Observation error variances are specified as in HR07, and are roughly similar to estimates by Crook (1996): $1.0 \, \text{K}^2$, $1.0 \times 10^{-6} \, \text{kg}^2 \, \text{kg}^{-2}$, and $2.0 \, \text{m}^2 \, \text{s}^{-2}$, for T_2, Q_2, and (U_{10}, V_{10}) respectively. The 30-minute surface and PBL profiles forecasts are verified at 0530 and 1730 UTC (0030 and 1230 LDT) against the surface observations and the available soundings, respectively, corresponding to assimilation cycling over 1.5 hours prior to verification. Skill scores, described in the next section, use 71 verification times for the surface variables and 57 and 65 verification times for the atmospheric profiles during the experiment at day and night time, respectively. The total number of verification times is determined simply by the number of available surface observations and the times that both the soundings and the 3D WRF forecast were available during the experimental period. The limited number of verification events raises the question of the statistical significance of our results, and to address it 90% CIs are calculated for each verification score following a bootstrapping technique detailed in the next section.

13.5 Verification metrics

The mean absolute error (MAE), rather than the root-mean-square error (RMSE), is chosen as a deterministic metric to quantify systematic error in the ensemble mean because it is more resistant to outliers (Jolliffe and Stephenson, 2003).

The usefulness of the probabilistic information provided by the SCM-EF system is measured by two main attributes of the probabilistic forecasts: reliability and resolution. Reliability is a measure of the statistical consistency between the predicted probability distributions and the verifying observations. In a reliable system the verifying observation is statistically identical to a random realization drawn from the predicted distribution. In other words, a reliable system provides unbiased estimates of the observed frequencies associated with different forecast probability values. Because reliability can be achieved by predicting the climatological probability distribution, it is necessary but not sufficient for a valuable probabilistic forecast. Resolution measures the capability of the forecast to distinguish between separate groups of observed events when these have a frequency different from the climatological frequency. Resolution and the closely related attribute of discrimination[2] are independent of reliability and represent a measure of the potential value of the system, since they are is insensitive to forecast bias.

The probablistic skill of the system is evaluated through the Brier skill score (BSS; Wilks, 1995) and the area under the relative operating characteristic (ROC) curve (AUR; Mason and Graham, 1999). Both of these metrics measure the system performance relative to a reference system. The BSS is easily decomposed into a reliability and a resolution term (Murphy, 1973) to understand the trade-offs in different components of probabilistic skill. The ROC curve has been widely used in the field of signal detection to distinguish between two alternative results (Mason, 1982). Thus, the AUR quantifies the ability of the forecast to discriminate between events. We define an "event" here to be a forecast value exceeding the 75th percentile of the observations. The climatology derived from the observations during the experiment period is used as reference in these calculations.

Since the number of realizations, I, is finite, an estimate of the uncertainty in the verification scores is required to allow meaningful statistical conclusions. The relevant question is: are the various scores obtained from our experiments under different model configurations statistically different, or are these differences insignificant? We quantify the scores' uncertainty through the estimation of CIs computed via a bootstrapping resampling procedure. This procedure consists of recalculation of the scores a number of times, N_b, with a sample of I realizations randomly extracted from the original dataset. We define CIs derived from the bootstrapped distributions of the scores using the "bias-correction and acceleration" (BCa) technique as described by Efron and Tibshirani (1993). The BCa intervals are corrections to the standard percentile intervals. For example, for a 90% CI the

---

[2] Resolution and discrimination are two closely related attributes but they are not the same: Resolution refers to the conditional distribution of the observations given the forecasts, while discrimination stands for the conditional distribution of the forecasts given the observations (Wilks, 1995).

interval would be (0.05, 0.95). The BCa technique adjusts this interval so that the mean of the bootstrapped distribution fits the original estimate of the score (from the original data set) and the width gives a more accurate estimate of the CI. CIs are calculated for the variables ef_{ijk}, thus they precisely represent the statistical significance of the effects of the factors relative to the base configuration.

13.6 Results and interpretation

Factor separation analysis has been applied to the metrics described in Section 13.5 to interpret the trade-offs between complexity and accuracy in the SCM-EF system. Verification is carried out in observation space (at the locations of the observations) and state space (SCM variables), thus investigating different aspects of the system. For example, whenever assimilation is activated, the verification at the surface quantifies the first-order ensemble performance and analyzes the set of variables that are directly corrected by the innovations ($\mathbf{y}^o - \mathbf{H}\mathbf{x}^b$) weighted by the estimated errors defined in Eqs. (13.5) and (13.6). In contrast, verification of the PBL profile analyzes model variables that are modified by the innovations and covariance with the state-space variables (see Anderson, 2003). Results for the full set of variables and verification times are presented and discussed in RH10 and a forthcoming paper. We present here results that illustrate the usefulness of the FS analysis in our study, and some of the most relevant conclusions about the relative effect of each of the studied model components.

13.6.1 Verification of the ensemble mean

Observation-space analysis

Figure 13.1 shows the analysis following the ASFSM in terms of the factor effect on the MAE, ef_{ijk} (Eqs. (13.7)–(13.14)), of 30-minute surface forecasts of T valid at 1730 and 0530 UTC (1230 and 0030 LDT, respectively). The error bars are 90% CIs derived from the bootstrapped distributions of ef_{ijk} values using the BCa technique and 1000 bootstrap samples. The MAE of a perfect forecast verified with error-free observations is zero, and a factor improving the system performance will reduce MAE. The variable ef_{000} (filled square) corresponds to the MAE of the basic model configuration (i.e., the system without activation of any of the three studied components, $ef_{000} = e_{000}$). When all the model components are included, the result is a reduced MAE denoted by the open square (total, e_{111}), both at day and night time. The system with all new components implemented shows MAE below the assigned observation error in the EF. CIs are largely reduced revealing that the improved system is less sensitive to specific weather conditions during the

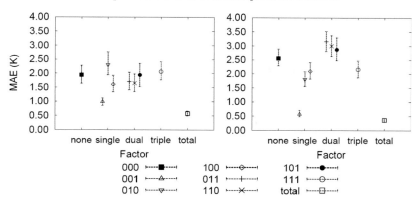

Figure 13.1 The system's performance as measured by the MAE of 30-minute surface forecasts of T valid at 1230 LDT (left panel) and at 0030 LDT (right panel). The MAE of the basic model configuration (e_{000}) and of the full model configuration (total, e_{111}) are denoted by filled and open squares respectively. The rest of the symbols represent the effect on the MAE (ef_{ijk}) resulting from each factor (as listed in Table 13.1) relative to the base configuration. Error bars represent 90% confidence intervals derived using the BCa bootstrapping technique.

BAMEX period. The rest of the symbols represent the effect of a specific (single or synergistic) factor on the system's performance as defined in Eqs. (13.8)–(13.14).

Pure assimilation (ef_{001}, triangle) shows the most significant contribution to reducing MAE and contracting CIs, particularly at night. The performance of the base configuration is poorer at night (0030 LDT) than during the day (1230 LDT). This is expected as PBL parameterizations (including MYJ) often fail under statically stable conditions (e.g., Mauritsen *et al.*, 2007, and references therein) when Monin–Obukhov similarity theory is invalid. Yet, pure assimilation is not the only factor that reduces MAE. Pure advection (ef_{100}, diamond) reduces MAE at day and night time too, but the magnitude of the effect is considerably smaller and CIs show that the effect may not always be observed. Pure parameterized radiation (ef_{010}, inverted triangle) shows opposite effects at day and night time. Clear improvement is attained at night showing its effectiveness in better simulating radiative cooling, but at day time it leads to a slight detrimental effect, which might result from the lack of cloud–radiation interaction when parameterized radiation was explicitly run within the SCM (i.e., the SCM does not explicitly consider clouds). In contrast, cloud–radiation interaction is accounted for when radiation is directly specified from the forcing 3D WRF forecasts, which include cloud processes. CIs show the clear dependence on weather scenarios (i.e., possibly clear skies vs. cloudy days). The SCM includes an option by which hydrometeor species from the forcing 3D WRF profiles can be advected into the SCM. This was not evaluated here, but it may diminish the detrimental effect.

The combined factors require further interpretation. Zero effect ($ef_{ijk} = e_{000}$) arising from a dual factor means that the individual factors combine linearly. A nonzero dual effect indicates a nonlinear, or synergistic, interaction between the analyzed individual components. A detrimental effect ($ef_{ijk} > e_{000}$) indicates that the combined action of the two factors partially (or completely) cancels the benefit achieved by one or both individual factors. On the contrary, a positive effect ($ef_{ijk} < e_{000}$) may be interpreted as the dual synergism canceling any individual detrimental effect to some extent or as the dual synergism further enhancing the positive effect achieved by the individual factors.

Figure 13.1 reveals that dual factors (+, x, and filled circle represent assimilation–radiation (ef_{011}), advection–radiation (ef_{110}), and advection–assimilation (ef_{101}) dual synergistic factors respectively) play different roles at day and night time. While at day time dual factors show a very weak positive or even null effect, the opposite occurs at night. Both pure radiation and advection play a stronger role at 0030 than 1230 LDT, leading also to somewhat more significant dual interactions at night. All single factors improve model performance at night, and dual nonlinear interactions among the three studied factors partially cancel the overall improvement. The same is true for the assimilation–radiation interaction. The triple synergism factor (open circle, ef_{111}) has a negligible effect at day time and a very slight positive effect at night.

Additional observation-space variables, i.e., U, V, and Q_v, show similar overall behavior (plots not shown here); e.g., assimilation plays the most significant role in reducing error and contracting CIs, bringing the performance of the system to levels in the range of assigned observation error in the filter, both at day and night time. Pure and nonlinear radiation interactions have no effect on the MAE of these variables (U, V, Q_v). Pure advection shows its most significant effect on U at day time. The remaining nonlinear interactions are very weak.

Similarly to Alpert *et al.* (1995), a general observation of the multi-factor results show that when the number of relevant factors increases, the role of any given factor is reduced because the synergistic interactions more often lead to some cancellation than positive synergism. One interpretation of this is that the system is near saturation in MAE, so that multiple components that improve MAE to a similar degree are not needed.

Analysis of the forecast profiles

The presentation of the effects of the three introduced system components on the MAE of forecast profiles is more complex than the analysis in observation space

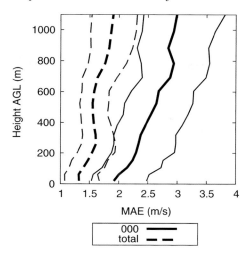

Figure 13.2 MAE of 30-minute forecasts of U profiles valid at 1230 LDT for the base configuration (bold continuous line, e_{000}) and for the full configuration including the three studied model components (bold dashed line, e_{111}). Thin lines are 90% confidence intervals calculated using the BCa bootstrapping technique.

because it includes multiple physical locations in the column. The complete analysis is presented in RH10, and here we focus on some of the most representative findings from factor separation analysis.

Figure 13.2 compares the MAE of 30-minute forecasts of U profiles valid at 1230 for the base ($ef_{000} = e_{000}$, solid bold lines) with the full system (total, e_{111}, dashed bold lines) configurations. Thin lines denote 90% CIs. The full configuration performs clearly better than the base configuration at all vertical levels. As above, the application of the ASFSM reveals that pure surface assimilation and advection are the most important factors for improving the system's performance. This might be expected because of the strong covariance between the surface and the column in the SCM. Radiation, also as expected, has no significant effect on U. Figure 13.3 shows the effect of three factors on the MAE of 30-minute forecasts of U profile valid at 1230 LT: (1) pure surface assimilation (left panel, solid bold lines represent the base configuration, $ef_{000} = e_{000}$; dashed bold lines are for ef_{001}); (2) pure advection (middle panel, line types as for left panel but dashed bold for ef_{100}); and (3) assimilation–advection synergism (right panel, line types as for left panel but dashed bold stands for ef_{101}). Thin lines denote 90% CIs. Surface assimilation and advection play similar roles at the first hundreds of meters AGL and advection dominates aloft (left and middle panels). This is caused by decreasing surface–atmosphere coupling (which determines the vertical extent of the effect of assimilation) with height. The assimilation–advection synergistic interaction gives

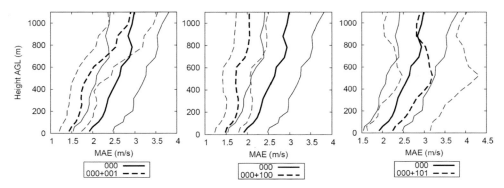

Figure 13.3 Same as Fig. 13.2 but dashed lines denote the resulting MAE when the effect of each of the following factors is added to the MAE of the base configuration: pure assimilation (ef_{001}) in left panel, pure advection (ef_{100}) in middle panel, assimilation–advection synergism (ef_{101}) in right panel.

rise to a detrimental effect (right panel) everywhere, indicated by the dashed curve lying to the right of the solid curve, which is most noticeable between 400 and 600 m AGL. At these intermediate levels the advection dominates the assimilation because the coupling between the observation and the column is weakening. Near 1000 m AGL, the assimilation does little and the synergism is small; the total benefit is simply from the advection. Comparison between Figs. 13.2 and 13.3 proves that the detrimental synergism results in less overall improvement than would be attained from the sum of pure assimilation and advection terms. However, 90% CIs partially overlap revealing that in many cases no net effect is observed (right panel Fig. 13.3).

Figures 13.4 and 13.5 present a similar analysis to that in Figs. 13.2 and 13.3 but for 30-minute forecasts of T profiles valid at 0030 LDT. The improvement achieved in the full configuration is clearly seen in Fig. 13.4, and once more the ASFSM facilitates simple interpretation. Radiation plays a minor role in the SCM compared to assimilation and advection (not shown here). Activation of the SCM radiation scheme improves the surface at night (Fig. 13.1), but the effect above the surface steeply decreases with height. Figure 13.5 reveals that the effect of surface assimilation dominates in the first $\sim$200 m AGL (left panel) and advection dominates above it (middle panel). The vertical extent of the effect of assimilation is shallower than that observed in Fig. 13.3, illustrating its dependence on the flow characteristics (convective unstable vs. nocturnal stable regime), which produce shallow covariance structures. The assimilation–advection nonlinear interaction (right panel) leads to a 90%-confident negative effect at levels between $\sim$50 and 400 m AGL. It should be noted that the local flow is characterized by shorter advection time scales above the PBL, which is shallow at night because of reduced

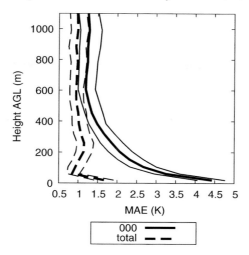

Figure 13.4 Same as Fig. 13.2 but for 30-minute forecasts of T profiles valid at 0030 LDT.

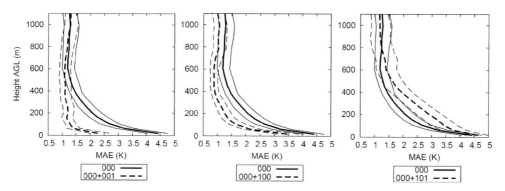

Figure 13.5 Same as Fig. 13.3 but for 30-minute forecasts of T profiles valid at 0030 LDT.

vertical momentum exchange. In contrast, very close to the surface (up to ∼50 m) the synergistic effect is null, which results from longer advection time scales and successful assimilation of surface observations.

13.6.2 *Probabilistic verification*

Observations-space analysis

As in the previous section we present selected findings that illustrate the usefulness of the ASFSM. We will limit this sub-section to the BSS and its decomposition. Figure 13.6 shows the factor effect (ef_{ijk}) on the BSS of 30-minute surface forecasts of Q_v valid at 1230 LDT (left panel) and at 0030 LDT (right panel). The

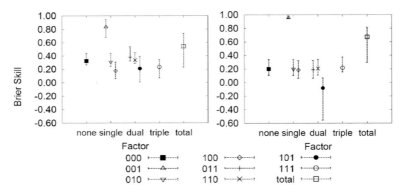

Figure 13.6 Same as Fig. 13.1 but for the Brier skill score (BSS) of 30-minute surface forecasts of Q_v.

BSS is positively oriented so that a perfect forecasting system gives BSS = 1, and BSS $\leq$ 0 indicates no skill or less skill than the reference. Figure 13.6 shows that even in its basic configuration ($ef_{000} = e_{000}$, filled square) the system shows probabilistic skill relative to the climatology of the observations, with somewhat better performance during day time. The addition of all new model components (total, e_{111}, open square) improves skill, particularly at night. Further inspection reveals that surface assimilation (ef_{001}, triangle) has the largest effect on probabilistic skill improvement, particularly at night (right panel) when it leads to a nearly maximum attainable value of BSS. As obtained for the MAE, surface assimilation proves to be the most efficient component in improving night-time system performance (right panel), which suffers from greater model error compared to day time.

Additional factors are responsible for slight negative effects at day time (left panel): pure advection (ef_{100}, diamond), assimilation–advection synergism (ef_{101}, filled circle), and to a small extent the triple assimilation–advection–radiation nonlinear interaction (ef_{111}, open circle). These partially cancel the improvement achieved through the assimilation of surface observations in the full configuration (total, e_{111}, open square). The width of the CIs is an indication of the dependence on the experiment conditions (i.e., the specific weather/flow regime). The effect of assimilation at night is characterized by narrow CIs indicating that surface assimilation improves performance under almost all conditions to the same extent. On the other hand, the effects of other factors are characterized by wide CIs, suggesting a stronger dependence on experiment conditions. Assimilation–advection synergism at night (right panel, ef_{101}, filled circle) exhibits the widest CIs, and a comparable behavior is observed at day time.

Examination of the reliability and resolution terms individually enables further interpretation. The decomposition is

$$BSS = (\text{resolution} - \text{reliability})/\text{uncertainty} \qquad (13.15)$$

Uncertainty here is purely a function of the distribution of observations and is therefore constant. Resolution less than or equal to the reliability indicates a prediction with no skill, and in a perfect system the resolution is equal to the sum of reliability and uncertainty. In general, the BSS is improved by maximizing the resolution (upper bound reliability plus uncertainty) and minimizing the reliability term (lower bound 0). Note again that resolution and reliability are independent because they result from different distributions.

Figures 13.7 and 13.8 complement the information provided in Fig. 13.6 and show the Brier reliability and resolution terms, respectively. The values in both figures follow the typical observed trend produced by present-day operational short-range weather forecasting systems, i.e., the reliability term is smaller than the resolution term (see, e.g., Jolliffe and Stephenson (2003)).

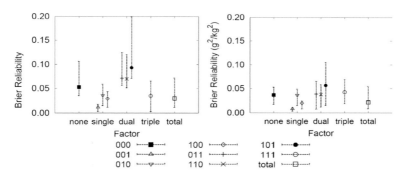

Figure 13.7 Same as Fig. 13.6 but for the Brier reliability term.

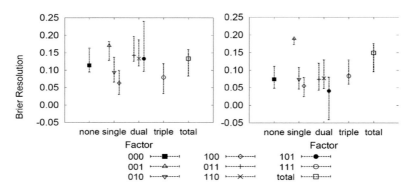

Figure 13.8 Same as Fig. 13.6 but for the Brier resolution term.

Figure 13.7 shows that in our experiments the full configuration (total, e_{111}, open square) is more reliable than the base configuration ($ef_{000} = e_{000}$, filled square). However, the calculated CIs reveal that this difference may not always be observed, particularly at day time. The ASFSM reveals the reasons for the little improvement observed in the full configuration. Pure surface assimilation (ef_{001}, triangle) is the most significant component in improving the system's reliability at day and night, consistent with its effect on the BSS (Fig. 13.6). Pure advection (ef_{100}, diamond) is the next most important component improving reliability, particularly at night (consistent with the BSS values). But the dual assimilation–advection synergism (ef_{101}, filled circle) shows again a destructive effect at day and night, agreeing with the BSS above.

Bias is a first-order contributor to reducing reliability. Positive effects can be interpreted as reducing bias in the forecast, which is certainly expected from including assimilation and advection. The detrimental nonlinear assimilation–advection term cancels once more part of the individual reliability improvements. Its wide CIs suggest that it strongly depends on the experiment conditions, e.g., the advection time scale as compared to the increment from assimilation, which may be largely determined by the weather/flow conditions. In contrast, narrow CIs on the positive contribution from surface assimilation prove the usefulness of data assimilation under a range of weather/flow conditions.

Figure 13.8 provides additional information to further account for the BSS behavior observed above. The full configuration (total, e_{111}, open square) exhibits better resolution than the base configuration ($ef_{000} = e_{000}$, filled square), but CIs show that at night this difference may not be statistically significant. Without further analysis of the effect of each factor we would conclude that, even though the full configuration includes several additional components, forecast resolution is in practice unaffected by them. During the day, surface assimilation (e f_{001}, triangle) is the only factor responsible for improving resolution, with very narrow CIs. In the full configuration this considerable improvement is partially canceled by the detrimental influences of pure advection (ef_{100}, diamond) and assimilation–advection synergism (ef_{101}, filled circle). The latter are accompanied once more by broad CIs.

Resolution is a measure of the capability of the system to distinguish between events. As such it is related to the sharpness of the prediction, which is a function of the ensemble spread. Assimilation contracts the spread of the system as it relaxes the ensemble members to the observations. Advection, as sampled from a distribution of WRF forecasts that is broader than the real distribution of analysis and forecast error (see Section 13.3), constitutes a factor that apparently produces too much spread in the ensemble, i.e., it destroys resolution. Parameterized radiation (ef_{010}, inverted triangle) shows under some scenarios a slight detrimental effect at night, but it is not significant.

Analysis of the forecast profiles

The probabilistic skill of the forecast profiles is analyzed through the calculation of the BSS and AUR. We do not intend to present here the full set of results, and instead summarize some of the most important findings. Extensive discussion will be presented in a forthcoming paper.

One of the most important findings from the complete analysis is the major role of surface assimilation in the probabilistic skill of the forecast profiles, with synergistic factors showing little effect. Figure 13.9 shows the BSS (left panel) and its decomposition into reliability and resolution terms (middle and right panels respectively) for 30-minute forecast of T profiles valid at 0030 LT for the base configuration ($ef_{000} = e_{000}$, solid bold lines) and the configuration including assimilation of surface observations only ($ef_{001} = e_{001}$, dashed bold line). Surface assimilation improves the BSS up to ~600 m AGL. It should be noted that the base configuration shows less skill (negative value of BSS) than the reference at levels closest to the ground (up to ~50 m). Assimilation provides skill and CIs indicate that in some cases it achieves its upper bound value (BSS = 1). Inspection reveals that the improvement due to assimilation results mostly from the enhanced Brier reliability term shown in the middle panel, while it shows no significant effect on the Brier resolution term (right panel).

We further inspect the effect of assimilation on the capability of the system to discriminate between groups of events through the calculation of the AUR. The AUR is a positively oriented metric, and values in the range [0.5, 1] imply discrimination. Figure 13.10 shows the AUR for 30-minute forecast of T profiles

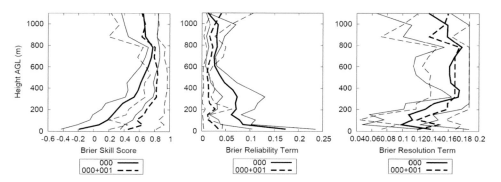

Figure 13.9 Brier skill score (BSS, left panel), Brier reliability term (middle panel) and Brier resolution term (right panel) (for the 75th percentile of the observations) of 30-minute forecasts of T profiles valid at 0030 LDT for the base configuration (bold continuous line, e_{000}) and for the configuration including assimilation only (bold dashed line, e_{001}). Thin lines are 90% confidence intervals calculated using the BCa bootstrapping technique.

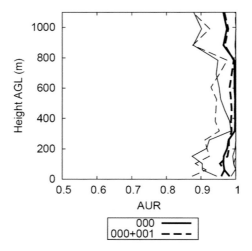

Figure 13.10 Same as Fig. 13.9 but for the area under the relative characteristic curve (AUR). The *x*-axis extends over the AUR skilful range.

valid at 0030 LT for the base configuration ($ef_{000} = e_{000}$, solid bold lines), and the configuration including only assimilation of surface observations ($ef_{001} = e_{001}$, dashed bold lines). Both the base system and the system including assimilation produce discriminating forecasts. This agrees with the Brier resolution term shown in Fig. 13.9, and confirms that the base system provides a useful probabilistic prediction system compared to climatology. Because the base system is nearly optimal in discrimination (AUR = 1) the assimilation cannot improve upon it.

13.7 Summary and conclusions

The present chapter summarizes our experience in applying the ASFSM to data assimilation and forecasting studies, and evaluating skill as opposed to a physical response. A system based on an SCM of the PBL and on assimilation of surface observations with an ensemble filter is used to produce probabilistic nowcasts of PBL profiles wherever surface observations are available. The SCM is simple but offers many components for implementation, and we seek to assess the value of some selected components on the system's performance with the aim of implementing the simplest efficient system configuration with good skill. Three system components are investigated: assimilation of surface observations, horizontal advective tendencies, and parameterized radiation. The performance of the system is investigated through deterministic and probabilistic verification metrics, inferring statistical significance of the verification scores through the calculation of CIs based on a bootstrapping technique.

The ASFSM applied to simultaneous verification of the surface and profile variables provides insight about the surface–atmosphere coupling in the system. For example, activation of the SCM radiation scheme improves surface nowcasts at night, but the effect above the surface steeply decreases with height. Vertical dependency is expected because radiative cooling is greatest near the surface, but ASFSM allows the skill improvements to be easily quantified. Assimilation, advection, and their mutual synergism play qualitatively similar roles at the surface and aloft. The significantly positive effect of assimilation at the surface and in the column variables is evidence of successful EF performance in two aspects: (a) correcting surface variables through the direct addition of the weighted innovations, and (b) correcting PBL profile variables through covariance of the observations with the profile. The vertical extent of the effect of assimilation on average decreases with height and is flow dependent, e.g., it is higher during day-time convective regimes than during night-time stable regimes. The effect of the advection tendencies depends on the advective-time scales, which are dictated by the characteristic flow at the various atmospheric levels and weather regimes. The intensity of the synergistic factors strongly depends on the coupling between the newly introduced system components through the model formulations, e.g., coupling of assimilation and advection tendencies through state augmentation. The results are helpful in determining whether these formulations and their implementation are useful or whether they require further improvement.

This study shows that without the use of the ASFSM it is hard to deduce unambiguous conclusions regarding the relative contribution of each of the studied model components. For instance, in some cases the simplest configuration (without including any additional model components) and the full configuration (including all three additional model components) lead to rather similar system performance. In these cases detailed Factor Separation (FS) analysis sheds light on the system behavior, and we discover that some components lead to significant improvement while others lead to a detrimental effect canceling part of the individually gained advantages.

Use of both deterministic and probabilistic verification procedures reveals that while some components have in both cases a positive effect others can lead to contradicting effects. Probabilistic verification shows that even the base configuration (without inclusion of any of the studied model components) has in most cases better probabilistic skill than the reference (observed climatology). In other words, the base system is probabilistically skilful. Assimilation of surface observations shows a consistent positive effect on model performance both deterministically (as proved by the MAE of the ensemble mean) and probabilistically (as proved by the BSS and its decomposition into reliability and resolution terms). Horizontal advection plays different roles depending on the performance metric. For instance, it significantly reduces MAE in the ensemble mean, but it can be harmful to ensemble resolution.

Parameterized radiation, which is the most computationally expensive among the three new components, exhibits a less significant effect than the advection and assimilation, both on the ensemble mean and ensemble distribution.

Synergistic effects often demonstrate that factors can easily cancel each other when the individual factors lead to enhancement of system performance. As interpreted by Alpert *et al.* (1995a) the importance of a specific component diminishes when additional components are included in the FS analysis. A further interpretation in the present context may be that system performance approaches saturation, so combining factors does not lead to measurable improvement matching the expected linear combination of improvements. However, synergism can lead to beneficial effects even when the individual components show no benefits. RH10 recently reported beneficial assimilation–radiation synergism for V-wind prediction when the radiation showed no individual effect.

One of the most important conclusions from this study is the fact that surface assimilation plays a more significant role in improving system performance than major model enhancements (e.g., parameterized radiation or advection). Horizontal advection is required to provide skilful ensemble-mean forecasts, and introduces the effects of 3D dynamics, but when acting simultaneously to assimilation it may cancel part of the improvement achieved through assimilation. Although state augmentation proved useful for tuning the horizontal advection tendencies (not shown here) we probably want to further improve its formulation or implementation in our system (e.g., more assimilation cycles may be required for state augmentation to become more effective).

The modeling strategy presented here will in turn be generalized to a 3D-assimilation and forecast system and the ASFSM will be one of our tools in the development effort.

Acknowledgements

The authors are grateful to the developers of the NCAR Data Assimilation Research Testbed (DART) for providing a useful platform for experimentation. We acknowledge M. Pocernich and E. Gilleland for guidance on the R-packages ("verification" and "bootstrap", www.r-project.org), which were used for our calculations. We acknowledge T. Hopson and Z. Klausner for insightful discussions on the bootstrap technique. D. Rostkier-Edelstein acknowledges the NCAR Research Applications Laboratory visitors program. Data were obtained from the Atmospheric Radiation Measurement (ARM) Program sponsored by the U.S. Department of Energy, Office of Science, Office of Biological and Environmental Research, Environmental Sciences Division.

14

Tagging systematic errors arising from different components of dynamics and physics in forecast models

T. N. Krishnamurti and Vinay Kumar

This study summarizes the results from a large number of forecast experiments with a global model with the objective of tagging errors that arise from different components of model dynamics and physics. To sort out such errors in non-linear systems is generally very difficult. In this study, we show examples of model forecasts that illustrate deficiencies (i.e., errors) that arise in the parameterization of cumulus convection and in the planetary boundary layer (PBL) physics. Our proposed Factor Separation method enables us to tag the systematic errors in our PBL formulation of moisture convergence. We note that the PBL scheme overestimates by roughly a factor of 2 the convergence near the tropical cloud base. These errors were the largest over meso-convective regions of deep convection. With such efforts at error tagging further improvements in modeling and forecasts can be achieved. We also address further work in this area of research.

14.1 Introduction

Factor Separation (FS), introduced by Stein and Alpert (1993) and Alpert *et al.* (2006), permits an explicit separation of atmospheric synergies among several salient features of a forecast model. They have exploited the FS method covering many scales of modeling to answer questions on issues such as the role of water and carbon dioxide for the understanding of global change. Their proposed method has many wide-ranging applications, as seen from the contents of this book. Our proposed study follows the same rationale in asking how one can sort out and tag errors that arise from different components of a non-linear model during the evolution of weather or climate.

Factor Separation in the Atmosphere: Applications and Future Prospects, ed. Pinhas Alpert and Tatiana Sholokhman. Published by Cambridge University Press. © Cambridge University Press 2011.

Our findings on the errors of a forecast model that arise from different components of model physics and dynamics are summarized and are extended here (Krishnamurti *et al.*, 1996; Krishnamurti *et al.*, 2004; O'Shay and Krishnamurti, 2004). This work parallels the concept of Alpert–Stein Factor Separation Methodology that was originated by Alpert (Stein and Alpert, 1993; Alpert, 1997; Krichak and Alpert, 2002). In a non-linear system, such as the atmosphere and the ocean, it is generally very difficult to pinpoint a source of error, since each component of model dynamics and physics interacts with each other continually in time and errors from one area compound those of other areas. It is, however, possible to use the data sets of model forecasts (collected for each time step of a model forecast) to describe the model tendencies for any specific model equation (such as the thermal equation) and seek the relative contributions to the model's total tendency from the component tendencies contributed by dynamics and physics. We carried out a multiple regression of the "observed" (based on frequent data assimilation) tendencies against the "model" component tendencies. The regression coefficients, called multipliers, for each of the terms of the model equation, thus obtained provide very useful information on the systematic error contributions for each model component. Such an exercise requires a statistical data sample based on the results from many forecasts. The multipliers are, in effect, obtained separately for each grid point in three dimensions for each model's prediction equation. Many models carry warm or cold biases, or they may carry moist or dry biases arising from different components of the model's dynamics and physics. This exercise can, in principle, map out the geographical and vertical distributions of such errors contributed by each of the dynamical and or thermodynamical terms of a given equation. Knowledge of such model biases can provide some guidance on areas where model improvements may be possible.

14.2 The multipliers that reduce forecast errors

Given the component and total tendency contributions of a model equation for a sequence of 30 days, we can estimate the total tendency errors from the relation

$$\varepsilon_{ijkl} = \left. \frac{\partial Q}{\partial t} \right)^{\text{mdl}}_{ijkl} - \left. \frac{\partial Q}{\partial t} \right)^{\text{analysis}}_{ijkl}$$

where i, j, and k denote an index for the three coordinates, and l denotes the variable; ε denotes the model error (i.e., model minus analysis estimates) and here mdl stands for model.

The model tendency may be written as a sum of the contributions from different terms:

$$\left. \frac{\partial Q}{\partial t} \right)^{\text{mdl}}_{ijkl} = \sum_m \left. \frac{\partial Q}{\partial t} \right)^{\text{mdl}}_{ijklm}$$

where m denotes various terms of the model equations. Hence the total tendency error can be represented as

$$\varepsilon_{ijkl} = \sum_m \left. \frac{\partial Q}{\partial t} \right)^{\text{mdl}}_{ijklm} - \left. \frac{\partial Q}{\partial t} \right)^{\text{analysis}}_{ijkl}$$

It is possible to define a three-dimensional (multiple regression based) multiplier λ_{ijklm} such that

$$\left. \frac{\partial Q}{\partial t} \right)^{\text{analysis}}_{ijkl} = \sum_m \lambda_{ijklm} \left. \frac{\partial Q}{\partial t} \right)^{\text{mdl}}_{ijklm} \quad \text{or}$$

$$0 = \sum_m \lambda_{ijklm} \left. \frac{\partial Q}{\partial t} \right)^{\text{mdl}}_{ijklm} - \left. \frac{\partial Q}{\partial t} \right)^{\text{analysis}}_{ijkl}$$

The determination of λ_{ijklm} utilizes the least-squares minimization procedure based on several multiple linear regressions. Here each of the coefficients is determined from month-long forecast data sets. The essential structure of λ_{ijklm} was found to be nearly invariant for two such separate computations. Once these λ_{ijklm} are determined, they provide a means for statistically corrected estimates of the forcing for the dynamics and physics for any of the equations while minimizing (toward zero) the total tendency error (we shall designate this as method-1 of this chapter).

The error at a grid location i, j, and k for a forcing A_{ijkl} is given by $(1 - \lambda_{ijkl})A_{ijkl}$. A_{ijkl} denotes the model value of a particular term in a forecast equation. These λs are determined for a large number of forecast experiments. These errors are four-dimensionally distributed. In the thermal equation λ conveys the message of warm or cold bias, and in the moisture equation it conveys the message of dry or wet biases of the model's moisture conservation equation for a given geographical location for any vertical level of the model.

14.3 Partitioning of errors

This is another method (we designate this as method-2) for exploring systematic errors that arise from different components of model physics and dynamics (Krishnamurti *et al.*, 2004). The strategy for extracting information on errors follows

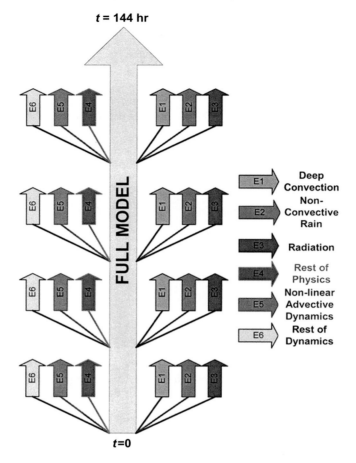

Figure 14.1 A schematic diagram showing the methodology for partitioning of model components in the framework of the "with and with" concept (Krishnamurti *et al.*, 2004).

a computational method schematically illustrated in Fig. 14.1 (Krishnamurti *et al.*, 2004). Here we run a control experiment that is based on a model that includes all its physics and dynamics. Other experiments started at time zero in each experiment suppress one feature of the model's dynamics (or physics) and make a one-time forecast. Thereafter, our experiment takes in the full fields from the control run and makes a next time step forecast suppressing that feature again. This process (of suppressing one area of physics or dynamics and retaining the full fields) is repeated till the end of a forecast run. What we have accomplished is that we have suppressed one area of physics or dynamics during each time step of this run, but the model recognizes the total tendency of all physics and dynamics at the start of each time step. This permits the excluded physics (or a dynamical component) to coexist with the rest of the evolving fields. This is called a "with and with" experiment, as contrasted with a "with and without" experiment.

An interesting example illustrates this methodology. A hurricane forecast was made suppressing all physics and frictional effects. Largely this was a steering flow experiment, where the hurricane moved from east to west in a three-day forecast experiment. That experiment was next repeated using a full physics experiment and the storm in the model recurved and eventually moved westwards. The second experiment demonstrated a close agreement with the observed track of that hurricane. An erroneous conclusion was drawn that dynamics moves the storm westwards, whereas the effects of dynamics plus physics recurves the storm. This is a typical "with and without" experiment, where we did not allow the effects of dynamics to coevolve with the physics in the first experiment. It is possible to determine the track from the dynamics and from dynamics plus physics using the "with and with" strategy illustrated in Fig. 14.1. Those results show that there is another dynamical contribution for the track forecast that coevolves with the physics and has far less of an error.

If the tendencies from a few of the physical processes are calculated for each time step and accumulated to the end the forecast, then we have a check sum, i.e., the total tendencies from those processes plus those from the rest of the model do equal the total model tendency. In spite of the non-linearity of the processes involved, and their interdependence, this check sum shows that unique contributions of a process can be separately identified. This is simply a book-keeping exercise and does not take into account how at adjacent time steps one process influences another process.

14.4 Short review of results for method-2

In a recent study, Krishnamurti *et al.* (1997) examined the simulation of a PNA pattern related to the ENSO. There have been many successful simulations of this pattern (Wallace and Gutzler, 1981; Nitta, 1986, 1987; and Lau, 1992). This is described as an external Rossby wave excited by the tropical heating of the ENSO over the central Pacific Ocean. The Rossby waves would be essentially non-divergent, away from the region of heating, e.g., Northern Pacific and North America, along the traverse of the PNA pattern. We noted a large increase of vorticity along the PNA lobes, from the analysis of the observed fields; hence, baroclinic processes must also be important.

After using a comprehensive global model, we successfully simulated the PNA pattern and subjected the computations to the aforementioned "with and with" strategy. In doing so we noted that the divergence term of the vorticity equation was in fact a major contributor for the evolution of the PNA pattern. We concluded

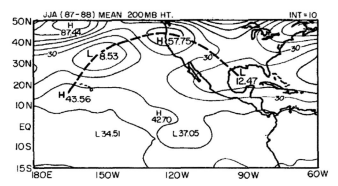

Figure 14.2 Seasonal prediction of the PNA pattern at 200 mb. This shows the differences in the 200 hPa geopotential 1987 minus 1988. The interval of analysis is 10 m.

that the PNA wave train was baroclinic and secondary heat sources and divergence along its traverse were important. This is the kind of finding that is possible from our proposed partitioning approach. The year 1987 was an El-Niño year, and the following year was not. A difference in the simulated seasonal prediction of the PNA pattern, as seen at the 200 hPa surface, is shown in Fig. 14.2. This clearly shows the strong PNA signal that was present in the simulations for the El-Niño year. In these experiments we noted that a stronger El-Niño pattern was contributed by the inclusion of heating and divergence terms in the forecast models outside of the tropics. That heating came from large-scale condensation and the divergence of the PNA components was part of the baroclinic dynamics. A simple barotropic model with an external heat source in the tropics over the central Pacific simulates a much weaker PNA pattern. The addition of baroclinicity and heating augments that pattern. These results were deduced from factor separations.

Using the "with and with" approach, O'Shay and Krishnamurti (2004) sorted out the physical and dynamical tracks for a hurricane forecast. They noted that the two processes are closely interlinked during a forecast, so that even steering, which appears largely dynamical, is continually affected by the evolving physics, especially the deep convective processes.

We have outlined two complementary methods for sorting out the contributions to a given forecast arising from the component physics or dynamics of that model. The former method is statistical and requires a data base generated from a large number of forecast experiments. The latter runs several forecasts, with the "with and with" strategy, using multiple processors and shares the information of a parent total forecast in order to evaluate the tendency from each component of dynamics or physics. We shall next illustrate the error in PBL modeling using the statistical approach.

14.5 Errors in the modeling of boundary layer fluxes (based on method-1)

Over the tropics, the modeling of the surface and PBL fluxes of moisture and their validation from observations is important for the simulations of the cloud layer and the dynamical feedbacks. Errors in the modeling of the PBL fluxes of moisture can impact large regions of the troposphere around these regions of strong moisture fluxes and convection. The validation of fluxes is very difficult since direct measurements of fluxes are not found over entire regions of numerical modeling. Such direct measures of fluxes are only met over regions of carefully designed field experiments, where special efforts are mounted to provide such measurements (Garstang, 1967). Over land areas of field experiments, boundary layer towers provide instruments that directly measure vertical motions and moisture, thus enabling the computation of moisture fluxes from the covariances of vertical velocity and humidity observations. Over oceans, such instrumentation can be mounted on balloons tethered to ships or placed in low-flying aircraft, and again can measure the vertical velocity and the humidity. Lacking such data sets, one can make use of forecast validations (as a function of space and time) of the large-scale humidity field in the PBL over a model domain. This validation is for grid-scale humidity fields, whereas the parameterization of the PBL relies on the sub-grid-scale moisture flux processes. This procedure provides an indirect validation for a PBL model that is followed in this study.

Here we shall illustrate the statistical method for determining the errors arising from the dynamical and physical terms of the moisture conservation equation as applied to the PBL of the FSU global model (Krishnamurti *et al.*, 2008). An outline of this model is provided in an appendix at the end of this chapter.

The moisture conservation equation in the PBL can be written in the form:

$$\frac{\partial q}{\partial t} = -V.\nabla q - \omega \frac{\partial q}{\partial p} + E - P - \frac{\partial}{\partial p}\overline{\omega' q'} \tag{14.1}$$

where q is the specific humidity. The local change of specific humidity is equated here to the horizontal advection of moisture, the vertical advection of moisture, evaporation, precipitation, and the sub-grid-scale vertical convergence of eddy flux of moisture (which carries the PBL physics). We shall be examining the errors of this term for our PBL parameterization. The PBL term denotes the vertical eddy flux convergence of moisture and is determined by the boundary layer theory, described later in this section. The height of the boundary layer is determined following our earlier study (Krishnamurti *et al.*, 2008). Presently, we have carried out as many as 111 forecast experiments and stored the values of these terms (i.e., terms of Eq. (14.1)) for each grid point in the horizontal and vertical (within the planetary boundary layer) at intervals of every three hours of the forecast. The averaged values of adjacent three-hourly forecasts of each term on the right hand side of Eq. (14.1) are stored at the mid time. The model tendency, i.e., the left hand side, is evaluated from the model-based values of moisture from the differences of these adjacent three-hourly intervals and the tendency also thus located at the mid time. This type of local moisture budget is exactly met by the model output only if one uses exactly the same numerics to evaluate each term as are in the design of the original model. That is not generally possible for post analysis, where one cannot use the entire features of the model, such as the semi-implicit time differencing and numerous other computational details. Thus, some computational residuals are always present, and the right hand side and the left hand side do not exactly balance. This is not a serious problem since we shall be statistically representing our results based on as many as 111 experiments, and in such a data base these errors tend to cancel out, providing a close balance between the right and the left hand sides of the above equation. The next step in this exercise is to use the following representation of moisture conservation, where the left hand side now denotes the observed three-hourly tendencies, based on data analysis:

$$\left.\frac{\partial q}{\partial t}\right)_{\text{analysis}} = \sum_i \lambda_i \times (\text{term } i \text{ of the model's right hand side})$$

Here the λ_i are unknown multipliers for the other parts of the above equation. If these λ_is are determined by a least-squares minimization using multiple regressions, against "observed" tendencies, then the terms multiplied by λ_is denote statistically improved values for these different terms. And $(1-\lambda_i)$ times that term denotes the systematic error for that term. This is expressed by

$$-(\lambda_i - 1)\frac{\partial}{\partial p}\,\overline{\omega' q'}$$

Since these λ_is vary from grid point to grid point in the horizontal as well as the vertical, it is possible to map the systematic errors of a given dynamical or physical forcing of a given model equation in three dimensions. Within the boundary layer, for the moisture equation, we are able to tell which terms are contributing to excessive moistening or drying. The diffusion term carries the details of the boundary layer physics for stable and unstable situations.

Since we use the σ coordinates, the tendency arising from the PBL physics is expressed by

$$\frac{\partial q}{\partial t} = \frac{g^2}{P_s^2}\frac{\partial}{\partial \sigma}\left(\rho^2 K \frac{\partial q}{\partial \sigma}\right) \tag{14.2}$$

where ρ is the air density, K is the eddy diffusivity, and q is the specific humidity.

This is a local-k closure approach in which the turbulent mixing is treated by a first-order closure. The eddy diffusivity coefficient K is defined as a function of the gradient Richardson number, R_i (Manobianco, 1988), i.e.,

$$K = l^2 \left| \frac{\sigma g}{RT} \frac{\partial \bar{V}}{\partial \sigma} \right| F(R_i) \tag{14.3}$$

where the mixing length l is given by

$$l = \frac{\kappa z}{l + \frac{\kappa z}{\lambda}} \tag{14.4}$$

The value of $\lambda = 150$ m for momentum and $\lambda = 450$ m for heat and moisture, z is the height of the σ surface, and κ is the von Karman constant ($= 0.35$); this follows Louis (1979). Prandtl surface-layer theory and observations showed that $l \cong \kappa z$ close to the surface. Also observations show that at greater elevations l might not continue to be proportional to distance from the ground. The Richardson number for a given atmospheric layer is given as

$$R_i = -\frac{\frac{l}{\theta} \frac{\partial \theta}{\partial \sigma}}{\left| \frac{\sigma}{RT} \frac{\partial \vec{V}}{\partial \sigma} \right|^2} \tag{14.5}$$

The stability function, $F(R_i)$, is of the form

$$F(R_i) = \begin{cases} \dfrac{1}{(1 + 5R_i)^2}, & \text{for stable conditions } (R_i \geqslant 0) \\[2ex] \dfrac{1 + \alpha \, |R_i|^{1/2} - 8R_i}{1 + \alpha \, |R_i|^{1/2}}, & \text{for unstable conditions } (R_i < 0) \end{cases} \tag{14.6}$$

where α is taken as 1.746 for momentum and 1.286 for heat and moisture.

Equations (14.3)–(14.6) define the formulation of the right hand side of Eq. (14.2). The FSU global model includes this formulation of the PBL. After obtaining the model output for the 111 experiments we store the tendencies for each grid location in these dimensions for the left and the right hand sides of the moisture tendency equation. We shall next address the PBL systematic error reduction by the proposed method.

14.6 Synoptic case history of maps and PBL

A three-day sequence during the northern summer season, July 1, 2, and 3 of the year 2007, is selected for this study. Although 111 forecasts were carried out to

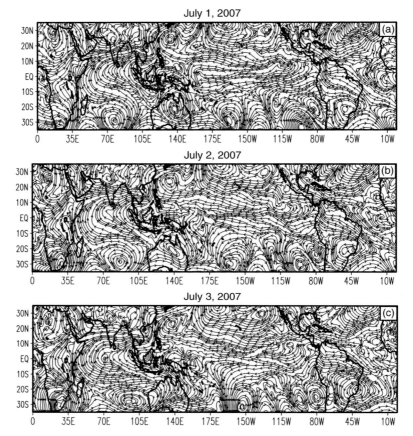

Figure 14.3 Maps of streamlines at 850 hPa for (a) 1 July 2007, (b) 2 July 2007, and (c) 3 July 2007.

obtain the statistics for the corrections of the PBL physics, the three selected days portray the kinds of corrections that are possible by this method. The synoptic situation is portrayed by the 850 hPa level flow fields for these days (Fig. 14.3). Here we note typical features such as the subtropical highs and the trade wind systems of the Atlantic and the Pacific oceans, the Inter Tropical Convergence Zone (ITCZ) of the same oceans and the southern hemisphere trades, the cross-equatorial flows, the southwest monsoon, the formation of a Bay of Bengal depression over its northern coast, and a plethora of frontal lows over the southern (winter) hemisphere along 30 °S. Our interest in this paper is to ask what kind of changes in the eddy convergence of moisture will result from the corrections of the systematic errors of the PBL physics.

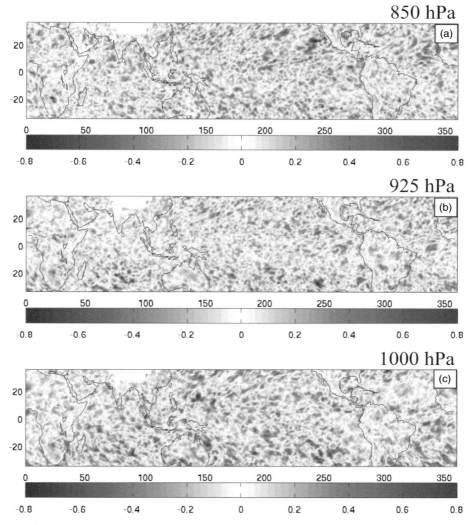

Figure 14.4 Spatial distribution of the weights (λ) calculated from multiple regression analysis with the 6 hr data from 23 March to 11 July 2007 at (a) 850 hPa, (b) 925 hPa, (c) 1000 hPa. Scale provided below each panel. Red shading denotes the positive values of weights while blue color denotes the negative values of weights. See plate section for color version.

In the next step of this exercise we replace the observed tendencies with the model tendencies to obtain the λ_is for the least-squares regression using data sets of 111 forecast experiments. Figure 14.4(a, b, c) illustrates the geographical distributions of λ for the PBL convergence of eddy flux of moisture at 850, 925, and 1000 hPa levels. Please note that these are close to the model σ surface at 0.850, 0.925, and 1.0 respectively. The red color denotes regions where the PBL convergence of fluxes of moisture by the model is overestimated and the blue color denotes regions where the convergence of flux of moisture is underestimated by the PBL parameterization. Overall we note a preponderance of red (overestimates) over the northern Pacific Ocean near 25° N and 20° S and blue color predominate over the equatorial Pacific Ocean. There are also some major regional differences in the structure of overestimates and underestimates as we examine vertical variation of λ. This suggests that the model output carries a complex structure that varies from one region to another and one must examine these overestimates (or underestimates) by the model in a regional phenomenological basis.

We shall next illustrate, Fig. 14.5, the fields of the PBL parameterization term (which is the convergence of eddy flux of moisture) for one-day forecasts valid on July 1, 2, and 3 of the year 2007 for three vertical levels, 850, 925, and 1000 hPa. When we examine the model output prior to correction then all of the values at the lowest level are positive, implying a convergence of flux of moisture by the PBL parameterization in the model. As we proceed to the two upper levels (850 and 925 hPa), all of the values are negative, implying a divergence of eddy flux of moisture. The largest magnitudes are found over the major rain belts, such as the ITCZ, the South Pacific Convergence Zone, southern (winter) hemisphere frontal zones, the eastern Bay of Bengal, and along the east coast of North America. The day to day variability of the convergence of flux is not very large. These are model-based values for day 1 of forecasts prior to the corrections. The choice for the selection of day 1 of forecasts is based on the decision to move past the spin-up time of model integration of the boundary layer physics to our initial data sets. We did note that the spin up was present in variables such as the sea-level pressure, precipitation, and vertical motions at individual grid points when the output at each time step at selected locations was examined. However, the spin up was only present in the first 12 hours. The changes after hour 12 seemed quite slow and monotonic, as can also be seen from the day to changes shown in Fig. 14.5. It is now possible to correct these fields for systematic errors by displaying the maps of λ_i times these fluxes. Note that λ varies along the three spatial coordinates.

With those corrections for the systematic errors for the overall tendency of moisture we arrived at Fig. 14.6, which includes nine similar panels. These are the corrected convergence of eddy flux of moisture for the three days at the three vertical levels. The picture changes somewhat; even at the lowest level we see some regions with divergence of eddy flux of moisture that were not present in the original model runs. The sign changes over regions of significant weather such as the ITCZ and the other weather systems listed above. The magnitudes also change quite a bit over these regions of significant weather. We note intense side lobes near regions of significant weather that carry opposite signs of eddy

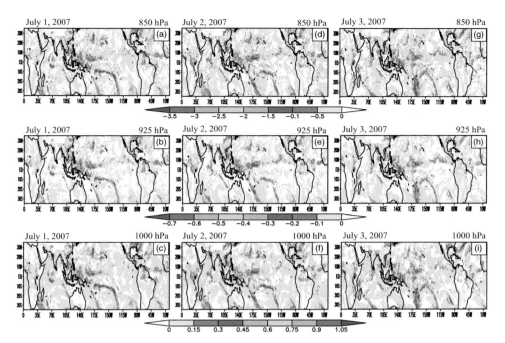

Figure 14.5 Diffusion term ($\times 10^{-4}$ s^{-1}) calculated from model forecasts at three vertical levels 850 hPa, 925 hPa, and 1000 hPa for (a)–(c) 1 July 2007, (d)–(f) 2 July 2007, and (g)–(i) 3 July 2007. See plate section for color version.

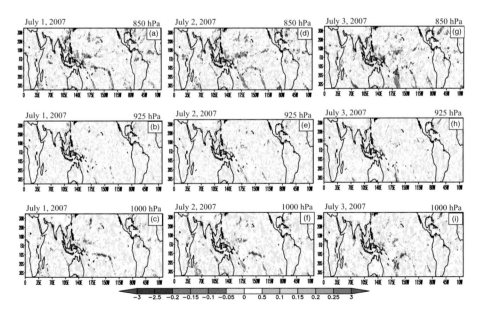

Figure 14.6 Same as in Fig. 14.5, except diffusion term is λ times the entries of Fig. 14.5. See plate section for color version.

convergence of flux of moisture to those found over the regions of significant weather. Even at the upper two levels we see some regions with positive signs, i.e., convergence of eddy flux of moisture is now present. Overall we note that over regions of significant weather, in the upper two levels, where the values on the divergence of flux from the PBL parameterization were corrected, the magnitudes have decreased somewhat, with additional side lobes carrying the opposite sign. The current parameterization, described in Section 14.5, overestimates the convergence/divergence of flux magnitudes somewhat. The correction scheme has reduced these magnitudes.

We shall next describe the changes that have been brought about by the proposed correction scheme within the PBL. The vertical cross sections of the convergence of fluxes in this local PBL scheme are considered here. It should be noted that this was one of the schemes with poor performance in terms of forecast skills (Krishnamurti *et al.*, 2008), and hence we expect large error corrections. Figures 14.7 and 14.8 carry these vertical cross sections of the eddy convergence of flux of moisture within the PBL over some significant weather events. These are the vertical structures of the convergence of eddy flux of moisture for the following four significant weather events regions: (i) Along the ITCZ over the western Pacific Ocean (between 120° E and 150° E , averaged between 5° S and 5° N); (ii) South Pacific Convergence Zone (between 170° E and 170° W and averaged 20° S to 30° S); (iii) ITCZ over the Atlantic Ocean (between 70° W and 10° W and averaged between 2° S and 7° N); and (iv) a south–north intersect from the Bay of Bengal into the land area of India (between 10° N and 25° N, averaged longitudinally between 88° E and 95° E). Each of these panels carries the vertical cross section from the model output and from the corrected field for these four synoptic regions. In the model output, prior to corrections the Pacific ITCZ region carries a weak convergence of flux below the 950 hPa level and largely has a few strong regions of divergence of eddy flux of moisture; the strongest such region is located near 144° E longitude where rainfall amounts were large. The corrected field called for a major reversal of sign over this part of the ITCZ, calling for a convergence of eddy flux of moisture. A weaker change of sign also was noted near 125° E, where again the corrections called for a convergence of eddy flux of moisture. In the active parts of the south Pacific convergence zone, between 171° E and 177° E longitudes, we also noted similar large corrections towards convergence of eddy flux of moisture. These changes are needed to account for the supply of moisture to the active convection above the PBL. That was a major deficiency of the scheme in heavy rain areas. The Atlantic ITCZ was not very active during the three days and the changes were more local in nature. The north–south intersect over the Bay of Bengal carried a large region of divergence of eddy flux of moisture between 17° N and 22° N above the 930 hPa level (see the figure). Those values were greatly reduced and even the sign

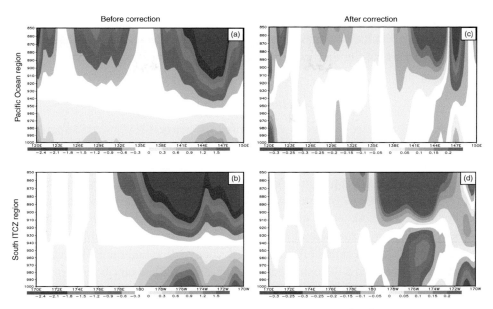

Figure 14.7 Vertical cross-sections of diffusion term ($\times 10^{-4}$ s^{-1}) over the major rain belt regions for 3 July 2007: (a) West Pacific Ocean region, averaged over 5° S to 5° N latitude belt; (b) South ITCZ region, averaged over 30° S to 20° S latitude belt; (c) same as (a) but after correction; (d) same as (b) but after correction. See plate section for color version.

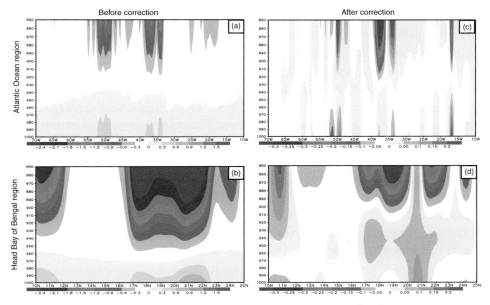

Figure 14.8 Same as in Fig. 14.7, except diffusion term ($\times 10^{-4}$ s^{-1}) is λ times the entries of Fig. 14.7. (a) Atlantic Ocean region, averaged over 2° S to 7° N latitude belt globe; (b) head Bay of Bengal region, averaged over 88° E to 95° E longitude belt; (c) same as (a) but after correction; (d) same as (b) but after correction. See plate section for color version.

corrected near 21° N latitude. Overall we see that the nature of these corrections over regions of active convection called for a net positive eddy flux convergence in the upper parts of the PBL near the cloud base.

14.7 Conclusions and future work

Our statistical approach requires the running of nearly 111 forecast experiments with a global model that enabled us to tag errors that arise from different components of model dynamics and physics. This follows the spirit of the factor separation method, introduced by Stein and Alpert (1993). To sort out such errors that arise in non-linear systems is generally very difficult. The use of large experimentation and statistical methods make such a factor separation possible to some extent. In this study, we show examples of model forecasts where we were able to illustrate deficiencies (i.e., errors) that arise in cumulus convection and the PBL parameterizations. We have used a diffusion-based, simple, local planetary boundary scheme that utilizes the mixing length theory and the bulk Richardson number based definition for the vertical exchange coefficient of diffusion. This is a well-known scheme that is being used in numerous weather and climate prediction models. In the context of the moisture conservation equation this diffusion measures the vertical eddy flux of moisture convergence. Krishnamurti *et al.* (2008) compared the performance of this scheme with respect to four other schemes for forecast error. Based on results for nearly 400 experiments they noted that this was in fact an inferior scheme in terms of resulting forecast errors. Errors in this formulation can provide erroneous values of moisture supply to the cloud base and can affect the cloud growth, especially in the tropics, leading to large forecast errors. Our proposed factor separation method enabled us to find the systematic errors in our PBL formulation of moisture convergence. We noted that the PBL scheme overestimates by roughly a factor of 2 the convergence near the tropical cloud base. These errors were the largest over meso-convective regions of deep convection. These are systematic errors that the PBL parameterization makes and such a systematic error was noted over the ITCZ region of the Pacific Ocean, over the South Pacific Convergence Zone and over a region where a Bay of Bengal monsoon depression forms. Over the region of the Atlantic Ocean ITCZ, where the structure of deep convection was not very sharp, these corrections were smaller.

Since the multipliers are statistically determined from 111 forecast experiments, they do provide a measure of the PBL errors in three dimensions. Within our PBL parameterization we can look for parameters that have somewhat questionable values. If those were multiplied by the λ_is (varying in three dimensions), then we could expect a correction to our formulation of the PBL physics. Looking through the PBL formulation, such expendable constants are in the values of α in Eq. (14.6), which were set to 1.286 for moisture transport. This can be adjusted over regions of extreme weather consistent with the λ_is. Altering those values over regions of deep convection, by multiplying those by the λ_is, may not encounter any singularities and can provide a major improvement for the PBL physics. We have also examined the λ_is for the other terms of the moisture tendency equation, such as the horizontal and vertical advection of moisture and the evaporation and precipitation. By far the most important corrections were called for by the eddy convergence of flux of moisture that is central to the PBL parameterization. There are several local and non-local PBL parameterization schemes (Manobianco, 1988; Stull, 1988; Kanamitsu, 1989; Holtslag and Boville, 1993; Hong and Pan, 1996; Basu *et al.*, 2002), and there are also some turbulent closure schemes (Mellor and Yamada, 1974, 1982; D'Alessio, 1998) It should be possible to find the systematic errors for those formulations and create an ensemble of forecast models that utilize such corrected schemes. With such efforts, further improvements in modeling and forecasts can be achieved: these are possible areas for future work.

Acknowledgments

The study was supported by the grant from NSF ATM-0636157.

Appendix: A brief description of FSU global spectral model

A detailed description of the global model used in this study can be found in Krishnamurti *et al.* (1991). The T170 version of the model, however, has been highly vectorized to reduce the model integration time. An outline of the model is as follows:

1. *The independent variables*: λ, θ, σ, t.
2. *The dependent variables*: Vorticity, divergence, temperature, and specific humidity.
3. *The horizontal resolution*: Triangular spectral truncation; T170 resolution has a 512×256 Gaussian transform grid with a horizontal separation of about 80 km at $20°$ latitude.
4. *The vertical resolution*: There are 27 layers in the vertical between 50 hPa and 1000 hPa. Model variables are staggered in the vertical using Charney–Phillips vertical discretization.
5. *The time integration scheme*: The divergence equation, thermodynamic equation, and pressure tendency equation are integrated implicitly while an explicit time integration scheme is used for the vorticity equation and moisture continuity equation. The tendencies of the physical processes are integrated using a forward time integration scheme.
6. *The space differencing scheme*: Spectral in the horizontal; centered differences in the vertical for all variables except moisture, which is handled by an upstream differencing scheme.

7. The surface topography is based on envelope orography (Wallace *et al.*, 1983).
8. *The parametrization of physical processes*: (a) Deep convection: based on the NCEP simplified Arakawa–Schubert cumulus parametrization scheme (Pan and Wu, 1995), with a saturated downdraft. Cloud ensemble is reduced to only cloud type with detrainment only from its top. It includes the effects of moisture detrainment from convective clouds, warming from environmental subsidence, and convective stabilization in balance with the large-scale destabilization rate; (b) shallow convection (Tiedke, 1984); (c) dry convection; (d) large-scale condensation (Kanamitsu, 1975). The scheme accounts for evaporation of falling precipitation; (e) PBL fluxes of heat, moisture, and momentum are calculated using similarity theory (Businger *et al.*, 1971); (f) vertical distribution of fluxes in the free atmosphere is based on a stability (Richardson number) dependent exchange coefficient (Louis, 1979); (g) fourth-order horizontal diffusion (Kanamitsu *et al.*, 1983); (h) long- and short-wave radiative fluxes are based on a band model and incorporate the radiative effects of water vapor, carbon dioxide, ozone, and clouds (Lacis and Hansen, 1974; Harshvardan and Corsetti, 1984); (i) parametrization of low, medium, and high clouds for radiative transfer calculation is based on threshold relative humidity; (j) surface temperatures: prescribed over the oceans, whereas over the land a surface energy balance coupled to the similarity theory determines the surface temperature including its diurnal cycle (Krishnamurti *et al.*, 1991).
9. *The non-linear normal mode initialization* (Kitade, 1983).
10. *Other boundary conditions*: Snow, monthly ($1° \times 1°$); SST, NCEP's $1° \times 1°$ weekly data; terrain, US Navy 5 minute data; vegetation, US Navy 5 minute data.

15

Some difficulties and prospects

P. Alpert and T. Sholokhman

This chapter discusses several difficulties as well as several prospects that are related to the FS methodology and that were not treated explicitly earlier in the book. First, the problem of dealing with a large number of factors is common to atmospheric research but becomes a serious computational burden when 2^n simulations are required. The most important point here is that we can limit our calculation to second-order interactions at the first stage and choose the most important factors, and only then continue toward a full factor separation.

Another somewhat related problem is how to deal with unchosen factors and with factor dependency. A further related problem is the fractional treatment that allows a finer analysis of potential non-linear effects of some factors. Particularly, how to reveal a significant variation in the effect of some factor due to change in a potential threshold.

A comparison of the FS against a different statistical factorial modeling (FM) method is performed. It is shown that the FS method has two clear advantages over the FM method. First, the sum of all 2^n contributions in an n-factor problem equals the full run result. This allows a closure of the separation as well as a percentage-wise analysis of all 2^n contributions that sum up to 100%. Second, the contribution by the zero- or basic-state is calculated and is one of the 2^n contributions. It is shown here and in many earlier FS studies that in the atmosphere the zero-state contribution can be significant and should be calculated.

The closure of the FS system and the basic-state contribution are important features in many problems and particularly in studying atmospheric developments.

Finally the FS perspective on human-caused climate change is highlighted.

Factor Separation in the Atmosphere: Applications and Future Prospects, ed. Pinhas Alpert and Tatiana Sholokhman. Published by Cambridge University Press. © Cambridge University Press 2011.

15.1 Large number of factors

A question often asked regarding the FS approach is how to treat a large number
of factors when the modeler cannot decide or choose which the important set of
factors is. This is particularly relevant to the method since the number of simulations
required increases with the exponent n where n factors are chosen. Stein and Alpert
(1993) (SA) addressed this question as follows. Regarding the number of relevant
factors in a particular problem, the investigator will frequently have a reasonably
good estimate for the dominant factors in a particular case. Basically, the FS could
be applied several times with varying factors. However, this involves a considerable
computational effort.

Another approach is when the decision on the chosen factors is difficult to make,
whether to include at the first stage a large number of potential factors but restrict
the interactions being resolved to the lower order ones. For instance, with ten
factors, instead of performing $2^{10} = 1024$ simulations, only $56 = (n, 0) + (n, 1) +$
$(n, 2)$ are required in order to obtain the double interactions only. At this stage, the
dominant factors can be selected based on up to double interactions only, and a
more complete FS may then be performed at the second stage. At this stage, with a
limited number of factors, all synergies or interactions can be computed. As shown
later by Alpert *et al.* (1995a) and other studies, often the higher-order interactions
become significantly small as the order increases, so using this methodology in
order to reduce the number of simulations right at the first stage may work well.

Another related question is about the applicability of the method once the useful
limit of predictability is reached. We believe that at this limit the method is probably
not meaningful because the errors will be large, and even larger with the higher-order
interactions where several simulations are analyzed in conjunction.

15.2 Unchosen factors and factor dependency

Alpert *et al.* (1995a) quantitatively illustrate that the contribution of any factor is not
unique and is strongly dependent on the particular set of factors (or corresponding
processes) being chosen by the investigator. The case chosen for illustration is an
impressive lee cyclogenesis event during the Alpine Experiment (ALPEX). Four
factors were chosen, i.e., topography (t), surface latent heat flux (l), surface sensible
heat flux (s) and the latent heat release (r). With four factors, eight different sets
of factors could have been chosen if one specific factor such as topography (t) is
included in each set. The potential eight sets are then {t}, {t, l}, {t, r}, {t, s}, {t, l, r},
{t, s, r}, {t, s, l}, and { t, s, l, r}. Notice that in this notation with { } we refer to
a potential numerical experiment for which the chosen factors or processes to be
tested are listed within these braces as a group. This notation should not be confused

with that of a specific contribution isolated for a particular experiment. For instance, the triple interaction *tlr* (as defined in Chapter 5) was isolated in an experiment for which four factors were chosen, that is, the experiment {t, s, l, r}. Of course, each set is a legitimate choice for a modeler who investigates the role of topography (*t*) in the lee cyclogenesis, and such examples are found in the literature (see SA). In Chapter 5 we have chosen the largest set with all four factors included, that is {t, s, l, r}. Alpert *et al.* (1995a, Figure 3) present the pure topographical contribution to the cyclone deepening for each of the eight possible sets of factors. It is not unexpected that the topographical contribution is largest when only one factor {t} was considered and the smallest – at least until hour 48 of the simulation was reached – when the largest set of factors {t, s, l, r} was chosen. Since, as the number of factors increases, the synergistic contributions between the new factors and topography are extracted from the original *t* contribution. In the extreme case where only topography was chosen as a factor, all synergistic contributions are tacitly assumed to be part of the topographical contribution, which makes the *t* contribution the largest. Although, in all eight experiments, the same contribution due to topography only is presented, the differences among the curves are quite significant. Basically, this implies that any sensitivity study of several factors is strongly dependent on the particular set being selected in that study. Since the different processes are in general not independent, the further increase in the number of factors diminishes the individual contribution of topography more and more, along with producing new synergistic terms associated with the new factors. Hence, the unavoidable introduction of synergism prohibits a quantitative comparison for the effects of several processes that is independent of the particular choice of factors. The more meaningful result may therefore be the calculation of variations in the contributions of a specific factor for several relevant groups of factors as suggested by Alpert *et al.* (1995a, Figure 3).

15.3 Fractional treatment

A fractional study of the FS approach was investigated by Krichak and Alpert (2002). The revised approach allows the determination of the role of the acting physical mechanisms as well as that of potential non-linearity of the modeling system responses. Application of the approach was demonstrated based on the simulations of an eastern Mediterranean weather development during 1–2 November 1994, performed with the Florida State University global spectral model. Thirteen model simulations with varying intensity of the turbulent surface fluxes in the runs were performed for the analysis. Two locations selected for sensitivity analysis of the model results represent the processes over the Red Sea and the Mediterranean Sea. It was shown that the variation of the FS results obtained in the experiments with different intensities of the factors may in some cases be significant. The degree of variability of the FS results obtained in the experiments with varying intensities of the factors under analysis provides useful information on sufficiency (or insufficiency) of available simulation results for an improved evaluation of the role of the factors acting in a meteorological process.

15.4 Comparison of the FS against a statistical factorial modeling method

A different approach for factor separation is a statistical one, also proposed with the 2^n factorial experiment (hereafter referred to as factorial modeling or FM for short). This tool is used in many scientific fields, including biotechnology experiments and numerical simulations, in order to evaluate the relative contribution of the entire set of possible interactions among the factors as compared to the contribution of each factor separately. The FM method/tool is briefly described next.

The 2^n factorial design of an experiment is discussed in a number of statistical textbooks. The experimental setup here is based on a single replication due to the fact that generally it is near to impossible to redo an experiment in atmospheric science in exactly the same conditions. The method and the notation follow Montgomery (2001).

This chapter focuses on special cases with n factors, each at only two levels, such as two values of temperature or pressure, or in other words the "high" and "low" levels of a factor, or perhaps the presence or absence of a factor. The statistical model for a 2^n design would include n single-factor effects, $\binom{n}{2}$ two-factor interactions, $\binom{n}{2}$ three-factor interactions, etc. until the n-factor interaction. For a 2^n design the complete model would contain $2^n - 1$ effects. Here we exemplify the 2^2 design.

For any two factors A, B, we denote the effect of a factor by a capital roman letter, following Montgomery (2001). Thus "A" refers to effect of factor A, "B" refers to the effect of factor B, and "AB" refers to the AB interaction.[1] In a 2^2 design the low and high levell of A and B are denoted by "$-$" and "$+$", respectively, on the A and B axes (Fig. 15.1). The four treatment combinations in the design are usually represented by lowercase letters (in the FS terminology these are referred to as the four basic runs). Thus a represents both A at the high level (+ in the figure) and the factor B at its lower level. Similarly, b represents A at the low level and B at its higher level. In addition, ab represents both factors at their high level (see Fig. 15.1). By the FM convention, (1) is used to denote both factors at their low level (which is f_0 in the FS terminology).

[1] Capital letters, i.e. A, B, in the FM terminology refer both to the factor itself and to the pure effect of this factor, while in FS methodology there are two different symbols, i.e., factor and "hat" symbols for net contribution as for instance, f and $\hat{f}$.

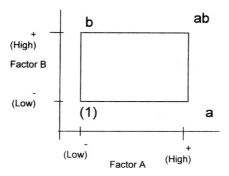

Figure 15.1 Treatment combinations in the 2×2 design.

The term "effect" (or "main effect" as defined by Montgomery (2001)) defines the average sensitivity of the result to a change in one factor and is calculated by subtracting the average of the results when this factor is set to its low value as compared to its high value. The "interaction" effect is the contribution of the combined changes by more than one factor within the experimental design, see ***ab*** in Fig. 15.1. The difference between the averages of the simulations with *A* being set at its high and low values, is calculated by

$$A = \frac{1}{2}\{[ab - b] + [a - (1)]\} = \frac{1}{2}[ab + a - b - (1)] \tag{15.1}$$

Similarly,

$$B = \frac{1}{2}\{[ab - a] + [b - (1)]\} = \frac{1}{2}[ab + b - a - (1)] \tag{15.2}$$

The effect of the interaction ***AB*** in the above design is therefore calculated by

$$AB = \frac{1}{2}\{[ab - b] - [a - (1)]\} = \frac{1}{2}[ab - a - b + (1)] \tag{15.3}$$

It should be noticed that in the FS terminology this is exactly the synergy contribution denoted as $\hat{f}_{12}$.

The "zero" effect in the FM approach becomes

$$I = \frac{1}{2}\{[ab + b] + [a + (1)]\} = \frac{1}{2}[ab + a + b + (1)] \tag{15.4}$$

It is interesting to note that in the FM method the "zero" effect is not identical to the run in which both factors are at their "low level" or switched off as in the FS terminology.

By graphical illustration it is easy to see the effects and the interactions in Fig. 15.1.

One advantage of the FM approach is that it allows further statistical analysis since it is based on a number of experiments. And therefore it involves a number of replications that can be separately analyzed by various statistical techniques. While the FS approach is based on a deterministic set of results, this may be eqvivalent to a single replication in the FM method.

Next, a simple analytical function is analyzed in order to illustrate some major differences between the FM and FS methods.

15.4.1 Case of a simple analytical function investigated by the FS and FM methods

The FS method approach: The simple bivariate polynomial was chosen:

$$f(x_1, x_2) = k_1 x_1 + k_2 x_2 + k_{12} x_1 x_2 + k_0 \tag{15.5}$$

First, the FS method is applied to the polynomial function.

It can easily be seen that the value of f and $\hat{f}$ in the different simulations are related through

$$
\begin{aligned}
f_0 &= k_0 & \hat{f}_0 &= k_0 \\
f_1 &= k_1 x_1 + k_0 & \hat{f}_1 &= k_1 x_1 \\
f_2 &= k_2 x_2 + k_0 & \hat{f}_2 &= k_2 x_2 \\
f_{12} &= k_1 x_1 + k_2 x_2 + k_{12} x_1 x_2 + k_0 & \hat{f}_{12} &= k_{12} x_1 x_2
\end{aligned}
\tag{15.6}
$$

For clarification, let use examine the substitution of zero and 1 for x_1 and x_2, which will represent the switch "off" and "on" situations, respectively:

$$x_1 = x_2 = 0 \quad \text{for the "off" case, and}$$
$$x_1 = x_2 = 1 \quad \text{for the "on" case.}$$

Hence, substituting in (15.6) yields

$$
\begin{aligned}
f_0 &= k_0 & \hat{f}_0 &= k_0 \\
f_1 &= k_1 + k_0 & \hat{f}_1 &= k_1 \\
f_2 &= k_2 + k_0 & \hat{f}_2 &= k_2 \\
f_{12} &= k_1 + k_2 + k_{12} + k_0 & \hat{f}_{12} &= k_{12}
\end{aligned}
\tag{15.7}
$$

As we can see, the sum of all four contribution is

$$\hat{f}_0 + \hat{f}_1 + \hat{f}_2 + \hat{f}_{12} = k_0 + k_1 + k_2 + k_{12} = f_{12} \tag{15.8}$$

where f_{12} is the final run or full result, which is equal to f_{12} at $x_1 = 1$, $x_2 = 1$, as expected.

In the FM approach, the bivariate polynomial is defined as

$$f(a,b) = k_1 a + k_2 b + k_{12} ab + k_0 \tag{15.9}$$

where a and b are the factors. The treatments and the four contributions, following (15.1)–(15.3), are

$$(1) = k_0 \qquad\qquad I = k_1 + k_2 + \frac{1}{2} k_{12} + 2k_0$$

$$a = k_1 + k_0 \qquad\qquad A = k_1 + \frac{1}{2} k_{12}$$

$$b = k_2 + k_0 \qquad\qquad B = k_2 + \frac{1}{2} k_{12} \tag{15.10}$$

$$ab = k_1 + k_2 + k_{12} + k_0 \qquad\qquad AB = \frac{1}{2} k_{12}$$

The sum of all contributions for the FM approach is, therefore,

$$I + A + B + AB = 2k_0 + 2k_1 + 2k_2 + 2k_{12} = 2ab \tag{15.11}$$

Notice that $2ab$ is just double the final result in the FS method. The sum of $2ab$ in the FS notation is $2f_{12}$ (see Eq. 15.8). This result is not in agreement with the intuitive expectation that the sum of all contributions should be exactly the result of the control run in which all factors are switched on. This is indeed the case for the FS approach.

In conclusion, the FS method has two clear advantages over the FM method. First, the sum of all 2^n contributions in an n-factor problem equals the control or the full run result. This allows a percentage-wise analysis of all 2^n contributions that sum up to 100%.

Second, the contribution by the zero or basic state is calculated and is one of the 2^n contributions. It was shown here and in many earlier FS studies that in the atmosphere the zero-state contribution can be significant. Since, when all chosen factors are switched off, there is often a significant contribution due to unchosen factors.

The closure of the FS system and the basic-state contribution are important features in many problems and particularly in studying atmospheric developments.

One attractive feature of the FM method is that it is designed to address a large number of experiments or replications, which is commonly the case in bio-technology and chemistry laboratory experiments; while the FS method was designed to deal in a deterministic, not statistical, approach. This means that a definite number of simulations defined by 2^n fully defines the 2^n possible contributions by *n* factors. The problem introduced by unchosen factors was studied at length by Alpert *et al.* (1995a) and Krichak and Alpert, (2002).

15.5 Factor separation perspective on human-caused climate change

Alpert *et al.* (2006) analyzed the potential synergies between the greenhouse gas (GHG) forcing(s) and water processes through different scales down to the leaf scale. Toward this goal, the FS methodology was employed.

Three experiments carried out by the authors all strongly suggest the existence of a significant CO_2–water synergy at all the involved scales. The experiments employed a very wide range of up-to-date atmospheric models that complement the physics currently introduced in most global circulation models (GCMs) for global climate change prediction.

Three modeling experiments that go from the small/microscale (leaf scale and soil moisture) to mesoscale (land-use change and CO_2 effects) and to global scale (greenhouse gases and cloudiness) all show that synergies between water and CO_2 are essential in predicting carbon assimilation, minimum daily temperature, and the global Earth temperature, respectively. The study also highlights the importance of including the physics associated with carbon–water synergy, which is mostly unresolved in global climate models, suggesting that significant carbon–water interactions are not incorporated or at least well parameterized in current climate models. Hence, there is a need for integrative climate models. As shown in earlier studies, the climate involves physical, chemical, and biological processes. To only include a subset of these processes limits the skill of local, regional, and global models to simulate the real climate system.

In addition, the results provide explicit determination of the direct and the interactive effect of the CO_2 on the terrestrial biosphere response. There is also an implicit scale-interactive effect that can be deduced from the multiscale effects discussed in the three examples. Processes at each scale – leaf, regional, and global – will all synergistically contribute to increase the feedbacks, which can decrease or increase the overall system's uncertainty depending on specific case/setup and need to be examined in future coupled, multiscale studies.

16

Summary

P. Alpert

Here, we first present the highlights from each of the applications in Chapters 3–15. Then, we summarize the most outstanding features of the FS method.

Following the introduction and description of the methodology in Chapters 1–2, this book provides a theoretical investigation of the FS method and defines common principles and analytical explorations of the factors and their synergies, Chapter 3. It should be noted that most of the studies employing the FS method did it in the numerical modeling framework, which simply does not allow an analytical investigation of the methodology. Hence, Chapter 3 gives the results of the method application with some basic and simple mathematical functions, which serve as the mathematical basis for investigation of more complicated processes.

Chapter 4 presents a paleoclimate model analysis that allows better understanding of the role of the temperature–albedo feedback, of the greenhouse gases (water vapor and CO_2), and of the insolation at the Last Glacial Maximum (LGM). The results are discussed using both the classical feedback analysis and the FS method. This helps to clarify the role the aforementioned factors play in the climate system. In particular, light is being shed on the seeming paradox that winters during mid-Holocene were milder than today in many regions of the Northern Hemisphere although the radiative forcing alone would cause a cooling.

In Chapter 5 we present three examples of factor separation on the medium scale in the atmosphere, often entitled meso-meteorology. In each of the three cases, some factors relevant to the specific problem were selected and special focus was given to the role played by the synergies as revealed by the FS approach.

The variable analyzed changes from the first case, in which it is the surface pressure at a cyclone center, to vorticity in the second case and to surface wind vectors over steep topography. As expected, independent of the variables chosen or of the

Factor Separation in the Atmosphere: Applications and Future Prospects, ed. Pinhas Alpert and Tatiana Sholokhman. Published by Cambridge University Press. © Cambridge University Press 2011.

factors' interactions, the synergies are shown to be the major players in mesoscale processes. In Chapter 6 the FS method is applied to assess the relative contribution of different factors to weather and climate processes. Model sensitivities to historical and future changes in land-use land-cover, biophysical and radiative effects of increased CO_2 concentration, and land-cover representation are assessed, emphasizing the importance of these climate surface factors when addressing regional climate change impacts.

Heavy precipitations and intense cyclones exert a high societal impact in the Mediterranean region and are investigated in Chapter 7. The roles exerted by boundary factors, such as the orography and surface heat fluxes, or by physical factors, such as the condensational latent heat release, have been isolated in numerous case studies.

In Chapter 8, the FS analysis is used to assess the significance of various land surface characteristics in the development and precipitation characteristics of convective storms occurring downwind of an urban region. In particular, the roles of topography, momentum fluxes, radiative heat fluxes, and latent and sensible heat fluxes are evaluated. The use of the FS technique in investigating the relative and interactive roles of different nucleating aerosols, including cloud condensation nuclei, giant cloud condensation nuclei, and ice nuclei, on the development, structure, and precipitation processes of tropical convection is also then described.

Chapter 9 focuses on the hydrodynamics of Lake Tanganyika. All year long, the thermocline of the lake oscillates about two equilibrium states. The FS method is employed to show that the forced oscillations of the thermocline are about twice as large as the free ones. Chapter 10 deals with industrial plumes and pollution effects on rainfall. The FS technique is applied to isolate the net contributions of waste heat, vapor, and CCN on the rainfall of a cumulus developing in the industrial plume. The contributions arising from mutual interactions among two or three factors are quite large and should not be neglected. Particularly, the synergistic interactions of the sensible heat and pollution effects contribute towards the accumulated rainfall.

Chapter 11 discusses crop simulation and land surface model-based assessments of the sensitivity to past and future changes in climatic conditions: increasing CO_2, soil moisture, temperature and radiative conditions, and crop management procedures (irrigation). The FS analysis shows that it is important to understand that biological responses are inherently dependent on multiple variables in the natural world and should not be limited to assessments of single specific parameters. In Chapter 12, application of the FS technique is demonstrated with a simplified problem that has an analytical solution – the Haurwitz sea-breeze (SB) linear model. The synergy of chosen factors is shown to explain well a paradoxical turning of the wind at low latitudes with low friction. Only the FS methodology provides a full picture of the wind evolution in this case.

Chapter 13 presents a forecasting application with data assimilation. The question addressed with the FS methodology is of finding an efficient system for probabilistic planetary boundary layer (PBL) very short-range forecasting (nowcasting) that can be employed when surface observations are present. Results show that the added complexity often improves the forecasts under most skill metrics, but that assimilation of surface observations is the most important contributor to improved skill. Chapter 14 also addresses a forecasting application employing a large number of forecast experiments with a global model, with the objective of tagging errors that arise from different components of model dynamics and physics. The proposed FS approach enables the tagging of systematic errors in PBL formulation of moisture convergence. With such efforts at error tagging, further improvements in modeling and forecasts can be achieved.

Chapter 15 summarizes some difficulties and prospects of the FS methodology that were not treated explicitly in other chapters of the book. First, the problem of dealing with a large number of factors, which is common to atmospheric research but becomes a serious computational burden when 2^n simulations are required. Another somewhat related problem is how to deal with unchosen factors and with factor dependency. A comparison of the FS methodology against a different statistical factorial modeling (FM) method is performed. The closure of the FS system and the basic-state contribution are important features in many problems, particularly in studying atmospheric developments.

Next, we summarize the most outstanding features of the FS method as revealed in our FS studies.

Sensitivity studies performed without separation of syngergistic terms may be misleading, as shown in several examples by Stein and Alpert (1993). In fact, prior to first publication in 1993, several attempts to quantify and compare the contributions from several factors may have led to wrong conclusions. These examples are all characterized by the following: (1) the study of at least two factors; (2) the performance of a number of experiments that switch off some of the factors; and (3) the estimation of the effect of certain factors by comparing various experiments.

Although the aforementioned interactions or synergies among factors were usually neglected, their significant role in some cases has been pointed out in several studies. The method presented in this book and its different applications, all the way from the large scale to the microscale, show a consistent and quite simple approach for isolating the resulting fields due to any interactions among factors, as well as those due to the pure factors, using linear combinations of a number of simulations.

It is now close to two decades since first publication of the basic methodology in "Factor separation in numerical simulations" and today this method provides a powerful tool for atmospheric research. A relatively large number of studies in recent decades have been devoted to various applications of this method in atmospheric research.

The Alpert–Stein Factor Separation Methodology can identify the most important factors and their combinations or synergisms. An additional problem that the FS method addresses is the stability of the chosen factor in the problem. This means, how the result reacts to small changes in the factors' values. The FS standard method works on the principle of "on-off"; in other words, the factors are one-by-one switched off (zeroed) and the intermediate result is investigated independently for every case. The case where all of the chosen factors are switched off is the basic case, which obviously does not depend on the influences of the factors or their combinations. The opposite case, with all of the factors "on", is the "control" or "full" result that includes all the factors and their synergistic contributions. The FS provides the methodology to distinguish between the pure influence of each and every factor as well as their mutual influence or synergies, which come into play when several factors, at least two, are "switched on" together. The understanding of which factor, or what combination of factors, is most significant for the final result is often very interesting in atmospheric studies. Discovering the most dominant factors in a specific problem can guide us to the important physical mechanisms and also to potential improvements in the model formulations.

Negative synergy in the atmosphere can be exemplified by the common case of saturation. In the atmosphere, for example, it can be most easily understood as humidity saturation for rainfall. In other words, processes in life are often not linear and the situation is similar or even more so in the atmosphere. Probably there are no factors in the real atmosphere that are not correlated at all. And this infinite chain of factor interactions, or some small subset of them, is commonly investigated by scientists, involving a finite number of assumptions.

One most attractive feature of the FS approach is the ability to quantify synergies or the interaction processes that were found to play a central role in many atmospheric processes as shown by several of the applications discussed in this book. There seems to be some basic psychological tendency in human thinking to linearization, resulting in the simplified assumption that synergies are small or can be ignored. Non-linearities in the atmosphere, however, are often quite significant and therefore need to be calculated and separated from the pure contributions of each factor.

Another attractive feature of the FS methodology is the fact that the system is well closed in the sense that the sum of all contributions, of the pure factors and their synergies, will always yield the result given by the control run in which all factors are switched on.

Some of the chapters include comparisons with other methodologies as well as methods similar to the FS method and discuss the advantages and disadvantages of the present method.

This is the first book to examine the FS method with several chosen examples. There are many other examples that were not incorporated, but the different chapters refer to numerous other FS studies.

Appendix

References employing the Alpert–Stein Factor Separation Methodology

Alpert, P., and Tsidulko, M. (1994). Project WIND: numerical simulations with Tel Aviv model PSU-NCAR model run at Tel Aviv University. In Mesoscale Modeling of the Atmosphere (Eds. R. A. Pielke and R. P. Pearce), Meteorological Monographs, 25, No. 47, Boston: American Meteorological Society, pp. 81–95.

Alpert, P., Tsidulko, M., and Stein, U. (1995). Can sensitivity studies yield absolute comparisons for the effect of several processes? *J. Atmos. Sci.*, 52, 597–601.

Alpert, P., Stein U., and Tsidulko, M. (1995). Role of sea fluxes and topography in eastern Mediterranean cyclogenesis. *The Global Atmosphere-Ocean System*, 3, 55–79.

Alpert, P., Krichak, S. O., Krishnamurti, T. N., Stein, U., and Tsidulko, M. (1996). The relative roles of lateral boundaries, initial conditions and topography in mesoscale simulations of lee cyclogenesis. *J. Appl. Meteor.*, **35**(7), 1091–1099.

Alpert, P., Tzidulko, M., Krichak, S. O., and Stein, U. (1996). A multi-stage evolution of an ALPEX cyclone. *Tellus*, **48A**, 209–220.

Alpert, P., Tzidulko, M., and Izigsohn, D. (1999). A shallow short- lived meso-beta cyclone over the gulf of Antalya, eastern Mediterranean. *Tellus*, **51A**, 249–262.

Alpert, P., Niyogi, D., Pielke, R. A. Sr., *et al.* (2006). Evidence for carbon dioxide and moisture interactions from the leaf cell up to global scales: Perspective on human-caused climate change. *Global and Planetary Change*, **54**, 202–208.

Berger, A. (2001). The role of CO_2, sea-level and vegetation during the Milankovitch-forced glacial-interglacial cycles. In *Proceedings "Geosphere-Biosphere Interactions and Climate"* (Eds. L. O. Bengtsson and, C. U. Hammer), New York: Cambridge University Press, pp. 119–146.

Carmona, I. (2004). The effect of aerosols on errors in weather and climate predictions. Ph.D. thesis, Faculty of Exact Sciences, Tel-Aviv University, Israel.

Castell, N., Stein, A. F., Salvador, R., Mantilla, E., and Millán, M. (2008). The impact of biogenic VOC emissions on photochemical ozone formation during a high ozone pollution episode in the Iberian Peninsula in the 2003 summer season. *Adv. Sci. Res.*, **2**, 9–15.

Claussen, M. (2009). Late Quaternary vegetation–climate feedbacks. *Clim. Past*, **5**, 203–216.

Claussen, M., Brovkin, V., and Ganopolski, A. (2001). Biogeophysical versus biogeo-chemical feedbacks of large-scale land cover change. *Geophys. Res. Lett.*, **28**, 1011–1014.

Crucifix, M., and Loutre, M. F. (2002). Transient simulations over the last interglacial period (126–115 kyr BP): feedback and forcing analysis. *Clim. Dyn.*, **19**, 417–433 (DOI 10.1007/s00382-002-0234-z).

Crucifix, M., Loutre, M.-F., Tulkens, P., Fichefet, T., and Berger, A. (2002). Climate evolution during the Holocene: A study with an Earth system model of intermediate complexity. *Clim. Dyn.*, **19**, 43–60.

Dallmeyer, A., Claussen, M., and Otto, J. (2010). Contribution of oceanic and vegetation feedbacks to Holocene climate change in monsoonal Asia. *Clim. Past*, **6**, 195–218.

Deng, L., and Lilly, D. K. (1992). Helicity effect on turbulent decay in a rotating frame. In *10th Symposium on Turbulence and Diffusion*, Portland, OR, American Meteorological Society, pp. 338–341.

De Ridder, K., and Galle, H. (1998). Land surface-induced regional climate change in southern Israel. *J. Appl. Meteorol.*, **37**, 1470–1485.

Deleersnijder, E., Ozer, J. and Tartinville, B. (1995). A methodology for model intercomparison : Preliminary results. *Ocean Modelling*, **107**, 6–9.

Eastman, J. L. (1999). Analysis of the effects of CO_2 and landscape change using a coupled plant and meteorological model. Ph.D. thesis, Colorado State University, USA.

Eastman, J. L., Coughenour, M. B., and Pielke, R. A. (2001). The effects of CO_2 and landscape change using a coupled plant and meteorological model. *Glob. Change Biol.*, **7**, 797–815.

Freitas, E. D., Silva Dias, P. L., Carvalho, V. S. B., *et al.* (2009). Factors involved in the formation and development of severe weather conditions over the megacity of São Paulo. *Eighth Symposium on the Urban Environment, Modeling and Forecasting in Urban Area*, Poster Session, 10–14 Jaunary 2009.

Fita, L., Romero, R., and Ramis, C. (2006). Intercomparison of intense cyclogenesis events over the Mediterranean basin based on baroclinic and diabatic influences. *Adv. Geosci.*, **7**, 333–342.

Gabusi, V., Pisoni, E., and Volta, M. (2008). Factors separation in air quality simulations, *Ecological Modeling*, **28**, 383–392.

Grossi, P., Thunis, P., Martilli A., and Clappier, A. (2000). Effect of sea-breeze on air pollution in the Greater Athens Area. Part II: Analysis of different emission scenarios. *J. Appl. Meteor.*, **39** (4), 563–575.

Guan, S., and Reuter, G. W. (1996). Numerical simulation of an industrial cumulus affected by heat, moisture and CCN released from an oil refinery. *J. Appl. Meteor.*, **35**, 1257–1264.

Hacker, J. P., Rostkier-Edelstein, D. (2008). PBL state estimation with surface observations, a column model, and an ensemble filter: probabilistic evaluation under various mesoscale regimes. *12th Conference on Integrated Observing and Assimilation Systems for Atmosphere, Oceans, and Land Surface (IOAS-AOLS)*, 20–24 January, 2008, New Orleans, USA.

Herot, A.-J., François, L., Brewer, S., and Munhoven, G. (2009). Impacts of land surface properties and atmospheric CO_2 on the Last Glacial Maximum climate: a factor separation analysis. *Clim. Past*, **5**, 183–202.

Homar, V., Ramis, C., Romero, R., *et al.* (1999). A case of convection development over the western Mediterranean Sea: A study through numerical simulations . *Meteor. Atmos. Phys.*, **71**, 169–188.

Homar, V., Romero, R., Stensrud, D. J., Ramis, C., and Alonso, S. (2003). Numerical diagnosis of a small, quasi-tropical cyclone over the Western Mediterranean: Dynamical vs. boundary factors. *Quart. J. R. Meteor. Soc.*, **129**, 1469–1490.

Horvath, K., Fita, L., Romero, R., and Ivancan-Picek, B. (2006) A numerical study of the first phase of a deep Mediterranean cyclone: Cyclogenesis in the lee of the Atlas mountains. *Meteor. Zeitschrift*, **15**, 133–146.

Horvath, K., Fita, L., Romero, R., Ivancan-Picek, B., and Stiperski, I. (2006). Cyclogenesis in the lee of the Atlas mountains: A factor separation numerical study. *Adv. Geosci.*, **7**, 327–331.

Jahn, A., Claussen, M., Ganopolski, A., and Brovkin, V. (2005) Quantifying the effect of vegetation dynamics on the climate of the Last Glacial Maximum. *Clim. Past Discuss.*, **1**, 1–16,

Khain, A. P., Rosenfeld, D., and Sednev, I. L. (1993). Coastal effects in the Eastern Mediterranean as seen from experiments using a cloud ensemble model with a detailed description of warm and ice microphysical processes. *Atmos. Res.*, **30**, 295–319.

Krichak, S. O., and Alpert, P. (1998). Role of the large-scale moist dynamics in Nov 1–5 1994 hazardous Mediterranean weather. *J. Geophy. Res.*, **103**(D16), 19453–19468.

Krichak, S. O., and Alpert, P. (2002). A fractional approach to the factor separation method. *J. Atmos. Sci.*, **59**, 2243–2252.

Krichak, S. O., Alpert, P., and Krishnamurti, T. N. (1997). Red Sea trough/cyclone development numerical investigation. *Meteor. Atmos. Phys.*, **63**, 159–170.

Lynn, B. H., Healy, R., and Druyan, L. M. (2009). Quantifying the sensitivity of simulated climate change to model configuration. *Clim. Change*, **92**, 275–298.

Lynn, B. H., Stauffer, D. R., Wetzel, P. J., *et al.* (2001). Improved simulation of Florida summer convection using the PLACE land model and a 1.5-order turbulence parameterization coupled to the Penn State–NCAR Mesoscale Model. *Mon. Wea. Rev.*,**129**(6),1441–1461.

Levy, G., Ek, M. (2001). The simulated response of the marine atmospheric boundary layer in the western Pacific warm pool region to surface flux forcing. *J. Geophy. Res.*, **106**(D7), 7229–7241.

Otto, J., Raddatz, T., Claussen, M., Brovkin, V., and Gayler, V. (2009). Separation of atmosphere-ocean-vegetation feedbacks and synergies for mid-Holocene climate. *Geophys. Res. Lett.*, **36**(9), L09701.

Pielke, R. A. Sr. (2002). Model sensitivity analyses. In *Mesoscale Meteorological Modeling*, 2nd Edition, San Diego, CA: Academic Press, p. 469.

Ramis, C., and Romero, R. (1995). A first numerical simulation of the development and structure of the sea breeze in the island of Mallorca. *Ann. Geophys.*, **13**, 981–994.

Ramis, C., Romero, R., Homar, V., Alonso, S., and Alarcón, M. (1998). Diagnosis and numerical simulation of a torrential precipitation event in Catalonia (Spain). *Meteorol. Atmos. Phys.*, **69**, 1–21.

Ramis, C., Romero, R., Homar, V., and Alonso, S. (2001). Lluvias torrenciales. *Investigación y Ciencia*, **296**, 60–68.

Reuter, G. W., and Guan, S. (1995). Effects of industrial pollution on cumulus convection and rain showers: A numerical study. *Atmos. Environ.*, **29**, 2467–2474.

Romero, R. (2008). A method for quantifying the impacts and interactions of potential vorticity anomalies in extratropical cyclones. *Quart. J. R. Meteor. Soc.*, **134**, 385–402.

Romero, R., Ramis, C., and Alonso, S. (1997). Numerical simulation of an extreme rainfall event in Catalonia: Role of orography and evaporation from the Sea. *Quart. J. R. Meteor. Soc.*, **123**, 537–559.

Romero, R., Ramis, C., Alonso, S., Doswell III, C. A., and Stensrud, D. J. (1998). Mesoscale model simulations of three heavy precipitation events in the western Mediterranean region. *Mon. Wea. Rev.*, **126**, 1859–1881.

References employing the factor separation methodology 253

Romero, R., Doswell III, C. A., and Ramis, C. (2000). Mesoscale numerical study of two cases of long-lived quasistationary convective systems over eastern Spain. Mon. Wea. Rev., 128, 3731–3751.
Rostkier-Edelstein, D., Hacker, J. P. (2009). Probabilistic nowcasting of PBL profiles with surface observations and an ensemble filter. In Proceedings of the 23rd Conference on Weather Analysis and Forecasting/19th, Conference on Numerical Weather Prediction, Omaha, NE, 1–5 June 2009.
Rostkier-Edelstein, D., and Hacker, J. P. (2010). The roles of surface-observation ensemble assimilation and model complexity for nowcasting of PBL profiles: A factor separation analysis. Wea. Forecast., (accepted).
Rozoff, C. M., Cotton, W. R, and Adegoke, J. O. (2003). Simulation of St. Louis, Missouri land use impacts on thunderstorms. J. Appl. Meteor., 42, 716–738.
Sholokhman, T., and Alpert, P. (2007). Factor separation in atmospheric modeling: A review. In Air, Water and Soil Quality Modeling for Risk and Impact Assessment (Eds. A. Ebel and T. L. Davitashvili), Berlin: Springer, pp. 165–180.
Sotillo, M. G., Ramis, C., Romero, R., Alonso, S., and Homar, V. (2003). Role of orography in the spatial distribution of precipitation over the Spanish Mediterranean zone. Clim. Res., 23, 247–261.
Stein, U. (1992). Physical mechanisms of cyclogenesis in the eastern Mediterranean. Ph.D. thesis, Tel-Aviv University.
Stein, U., and Alpert, P. (1993). Factor separation in numerical simulations. J. Atmos. Sci., 50, 2107–2115.
Takle, E., Segal, M., and Pan, Z. (2006–2009). Evaluating effects of climate changes on Midwest agroecosystems using a climate-crop coupled model, project DOE/NICCR Midwestern Center.
Thunis, P., and Cuvelier, C. (2000). Impact of biogenic emissions on ozone formation in the Mediterranean area: a BEMA modelling study. Atmos. Environ., 34, 467–481.
Tsidulko, M. (1997). A numerical investigation of synergic contributions to lee cyclogenesis. Ph.D. thesis, Tel-Aviv University.
Tsidulko, M., and Alpert, P.(2001). Synergism of upper-level potential vorticity and mountains in Genoa lee cyclogenesis: A numerical study. Meteor. Atmos. Phys., 78(3/4), 261–285.
van den Heever, S. C., Carrio, G. G., Cotton, W. R., DeMott, P. J., and Prenni, A. J. (2006). Impacts of nucleating aerosol on Florida storms. Part I: Mesoscale simulations. J. Atmos. Sci., 1752–1775.
Yin, Q. Z., and Berger, A. (2010). Insolation and CO_2 contribution to the interglacial climate before and after the Mid-Brunhes Event. Nature Geosci., 3, 243–246.

References

Ackerman, S. A., and Knox, J. (2002). *Meteorology: Understanding the Atmosphere*, Pacific Grove, CA: Brooks/Cole.

Alagarswamy, G., Boote, K. J., Allen, L. H. Jr., and Jones, J. W. (2006). Evaluating the CROPGRO–Soybean Model ability to simulate photosynthesis response to carbon dioxide levels. *Agron. J.*, **98**, 34–42.

Alapaty, K., Pleim, J., Raman, S., Niyogi, D. S., and Byun, D. (1997). Simulation of atmospheric boundary layer processes using local- and nonlocal closure schemes. *J. Appl. Meteorol.*, **36**, 214–233.

Alapaty, K., Seaman, N., Niyogi, D. S., and Hanna, A. (2001). A technique for assimilating surface data to improve the accuracy of atmospheric boundary layer simulations: A 1-d model study. *J. Appl. Meteorol.*, **40**, 2068–2082.

Albrecht, B. (1989). Aerosols, cloud microphysics, and fractional cloudiness. *Science*, **245**, 1227–1230.

Allen Jr., L. H., and Boote, K. J. (2000). Crop ecosystem responses to climatic change: soybean. In K. R. Reddy and H. F. Hodges (eds.), *Climate Change and Global Crop Productivity*, New York: CAB International Publishing, pp. 133–160.

Allen Jr., L. H., Boote, K. J., Jones, J. W., *et al.* (1987). Response of vegetation to rising carbon dioxide: Photosynthesis, biomass, and seed yield of soybean. *Glob. Biogeochem. Cycles*, **1**, 1–14.

Allen Jr., L. H., Bisbal, E. C., and Boote, K. J. (2004). Nonstructural carbohydrates of soybean plants grown in subambient and superambient levels of CO_2. *Photosynth. Res.*, **56**, 143–155.

Alpert, P. (1997). Factor separation: an emerging tool in the investigation of Mediterranean cyclogenesis. *Symposium on Cyclones and Hazardous Weather in the Mediterranean INM/WMO*, Palma, Spain, pp. 653–656.

Alpert, P., and Neumann, J. (1984) On the enhanced smoothing over topography in some mesometeorological models. *Bound. Layer Meteor.*, **30**, 293–312.

Alpert, P., and Tsidulko, M. (1994). Project WIND: numerical simulations with Tel-Aviv PSU/NCAR model at Tel-Aviv University 1994. In *Mesoscale Modeling of the Atmosphere*, Meteorological Monograph, No. 25, American Meteorological Society, pp. 81–95.

Alpert. P., Stein, U., Tsidulko, M., and Neeman, B. U. (1993). Synergism in weather and climate (unpublished).

Alpert P., Tsidulko, M. and Stein, U. (1995a). Can sensitivity studies yield absolute comparisons for the effect of several processes? *J. Atmos. Sci.*, **52**, 597–601.

Alpert, P., Stein, U., and Tsidulko, M. (1995b). Role of sea fluxes and topography in eastern Mediterranean cyclogenesis. *Glob. Atmos. Ocean System*, **3**, 55–79.

Alpert, P., Krichak, S. O., Krishnamurti, T. N., Stein, U., and Tsidulko, M. (1996a). The relative roles of lateral boundaries, initial conditions and topography in mesoscale simulations of lee cyclogenesis. *J. Appl. Meteor.*, **35**(7), 1091–1099.

Alpert, P., Tzidulko, M., Krichak, S. and Stein, U. (1996b). A multi-stage evolution of an ALPEX cyclone. *Tellus*, **48A**, 209–220.

Alpert, P., Tzidulko, M., and Izigsohn, D. (1999). A shallow short- lived meso-beta cyclone over the gulf of Antalya, eastern Mediterranean. *Tellus*, **51A**, 249–262.

Alpert, P., Niyogi, D., Pielke Sr., R. A., *et al.* (2006). Evidence for carbon dioxide and moisture interactions from the leaf cell up to global scales: Perspective on human-caused climate change. *Global Planet. Change*, **54**, 202–208.

Anderson, J. L. (2001). An ensemble adjustment Kalman filter for data assimilation. *Mon. Wea. Rev.*, **129**, 2884–2903.

Anderson, J. L. (2003). A local least squares framework for ensemble filtering. *Mon. Wea. Rev.*, **131**, 634–642.

Anderson, J. L. (2007). Exploring the need for localization in ensemble data assimilation using a hierarchical ensemble filter. *Physica D*, **230**, 99–111.

Andreae, M. O., Rosenfeld, D., Artaxo, P. *et al.* (200) Smoking rain clouds over the Amazon. *Science*, **303**, 1337–1342.

Antenucci, J. P. (2005). Comment on "Are there internal Kelvin waves in Lake Tanganyika?". *Geophys. Res. Lett.*, **32**(22), L22601, doi: 10.1029/2005GL024403.

Antenucci, J. P., and Imberger, J. (2001). Energetics of long internal gravity waves in large lakes. *Limnol. Oceanogr.*, **46**(7),1760–1773.

Arora, V. K. (2002). Modeling vegetation as a dynamic component in soil-vegetation-atmosphere-transfer schemes and hydrological models. *Rev. Geophys.*, **40**, doi:100610.1029/2001RG000103.

Auer, A. H., Jr. (1976). Observations of an industrial cumulus. *J. Appl. Meteor.*, **15**, 406–413.

Baidya Roy, S., Hurtt, G. C., Weaver, C. P., and Pacala, S. W. (2003). Impact of historical land cover change on the July climate of the United States. *J. Geophys. Res.* **108**(D24), 4793, doi:10.1029/2003JD003565.

Ball, J., Woodrow, I., and Berry, J. (1987). A model predicting stomatal conductance and its contribution to the control of photosynthesis under different environmental conditions. In *Progress in Photosynthesis Research*, Vol. IV, Dordrecht: Martinus Nijhoff, pp. 221–224.

Balzano, A. (1998). Evaluation of methods for numerical simulation of wetting and drying in shallow water flow models. *Coastal Engineering*, **34**(1–2), 83–107.

Basu, S., Iyengar, G. R., and Mitra, A. K. (2002). Impact of a nonlocal closure scheme in a simulation of a monsoon system over India. *Mon. Wea. Rev.*, **130**, 161–170.

Beltrán, A. (2005). Using a coupled biospheric-atmospheric modeling system (GEMRAMS) to model the effects of land-use/land-cover changes on the near surface atmosphere. Ph.D. Dissertation, 186 pp. Colorado State University.

Beltrán-Przekurat, A., Marshall, C. A., and Pielke Sr., R. A. (2008). Ensemble re-forecasts of recent warm-season weather: Impacts of a dynamic vegetation parameterization. *J. Geophys. Res. Atmos.*, **113**, D24116, doi:10.1029/2007JD009480.

Benjamin, S. G., and Carlson, T. N. (1986). Some effects of surface heating and topography on the regional severe storm environment. Part 1: Three-dimensional simulations. *Mon. Wea. Rev.*, **114**, 307–329.

Berg, W., L'Ecuyer, T., and van den Heever, S. C. (2008). Evidence for the impact of aerosols on the onset and microphysical properties of rainfall from a combination of active and passive satellite sensors. *J. Geophy. Res.*, **113**, D14S23, doi:10.1029/2007JD009649.

Berger, A. (1978). Long-term variations of daily insolation and Quaternary Climatic Changes. *J.Atmos. Sci.*, **35**(12), 2362–2367.

Berger, A. (2001). The role of CO_2, sea-level and vegetation during the Milankovitch-forced glacial-interglacial cycles. In L. O. Bengtsson and C. U. Hammer (eds.), *Proceedings Geosphere-Biosphere Interactions and Climate*. New York: Cambridge University Press.

Berger, A., and. Tricot, Ch. (1992). The greenhouse effect. *Surv. Geophys.*, **13**, 523–549.

Berger, A., Tricot, Ch., Gallée, H., and Loutre, M. F. (1993). Water vapour, CO_2 and insolation over the last glacial-interglacial cycles. Meeting on "Palaeoclimates and Their Modelling with Special Reference to the Mesozoic Era", Royal Society of London, *Phil. Trans. R. Soc. Lond. B*, **341**, 253–261.

Berry, E. X., and Reinhardt, R. L. (1973). *Modeling of Condensation and Collection Within Clouds*. Desert Research Institute Physical Science Publication, 16, University of Nevada, Reno.

Bleck, R., and Mattocks, C. (1984). A preliminary analysis of the role of potential vorticity in Alpine lee cyclogenesis. *Beitr. Phys. Atmos.*, **57**, 357–368.

Booker, F. L., Miller, J., Fiscus, E. L., Pursley, W. A., and Stefanski, L. A. (2005). Comparative responses of container versus ground-grown soybean to elevated carbon dioxide and ozone. *Crop. Sci.*, **45**, 883–895.

Boone, A., Calvet, J.-C., and Noilhan, J. (1999). Inclusion of a third soil layer in a land surface scheme using the force-restore method. *J. Appl. Meteorol.* **38**, 1611–1630.

Boote, K. J., Jones, J. W., Hoogenboom, G., and Wilkerson, G. G. (1997). Evaluation of the CROPGRO-Soybean model over a wide range of experiments. In M. J. Kropff, P. S. Teng, P. K. Agrawal, *et al.* (eds.), *Applications of System Approaches at Field Level*. Dordrecht, the Netherlands: Kluwer, pp. 113–133.

Boote, K. J., Jones, J. W., and Hoogenboom, G. (1998). Simulation of crop growth: CROP-GRO. In R. M. Peart and R. B. Curry (eds.) *Agricultural Systems Modeling and Simulation*. New York: Marcel Dekker, pp. 651–692.

Borys, R. D., Lowenthal, D. H., Wetzel, M. A., *et al.* (1998). Chemical and microphysical properties of marine stratiform cloud in the North Atlantic. *J. Geophys. Res.*, **103**, 22,073–22,085.

Box, G. E. P., Hunter, J. S., and Hunter, W. G. (1978). *Statistics for Experimenters: An Introduction to Design, Data Analysis, and Model Building*. New York: Wiley.

Braham, R. R., Semonin, R. G., Auer, A. H., Changnon, S. A., and Hales, J. M. (1981). Summary of urban effects on clouds and rain. In *METROMEX: A Review and Summary*, Meteorological Monograph, No.18, American Meteorological Society, pp. 141–152.

Brown, R. A., and Rosenberg, N. J. (1997). Sensitivity of crop yield and water use to change in a range of climatic factors and CO_2 concentrations: A simulation study applying EPIC to the central USA. *Agric. For. Meteorol.*, **83**, 171–203.

Budyko, M. I. (969). The effect of solar radiation variations on the climate of the earth. *Tellus*, **21**, 611–619,.

Businger, J. A., Wyngaard, J. C., Izumi, Y., and Bradley, E. F. (1971). Flux profile relationship in the atmospheric surface layer. *J. Atmos. Sci.*, **28**, 181–189.

Buzzi, A., and Tibaldi, S. (1978). Cyclogenesis in the lee of the Alps: Case study. *Quart. J. Roy. Meteor. Soc.*, **1**, 271–287

Buzzi, A., Trevisan, A., and Tossi, E. (1985). Isentropic analysis of a case of Alpine cyclogenesis. *Beitr. Phys. Atmos.*, **58**, 273–284.

Buzzi, A., Trevisan, A., Tibaldi, S. and Tossi, E. (1987). A unified theory of orographic influences upon cyclogenesis. *Meteor. Atmos. Phys.*, **36**, 1–107.

Cabrera, A. L., and Willink, A. (1980). *Biogeografía de América Latina*. 2nd edn. Secretaría General de la Organización de los Estados Americanos, Programa Regional de Desarrollo Científico y Tecnológico.

Cao, L., Bala, G., Caldeira, K., Nemani, R., and Ban-Weiss, G. (2009). Climate response to physiological forcing of carbon dioxide simulated by the coupled Community Atmosphere Model (CAM3.1) and Community Land Model (CLM3.0). *Geophys. Res. Lett.*, **36**, L10402, doi:10.1029/2009GL037724.

Carlson, T. N., and Bunce, J. A. (1996). Will a doubling of atmospheric carbon dioxide concentration lead to an increase or a decrease in water consumption by crops? *Ecol. Model.*, **88**, 241–246.

Castro, C. L., Pielke Sr., R. A. and Leoncini, G. (2005). Dynamical downscaling: Assessment of value retained and added using the Regional Atmospheric Modeling System (RAMS). *J. Geophys. Res.*, **110**, D05108, doi:10.1029/2004JDD004721.

Chang, C. B., Perkey, D. J., and Kreitzberg, C. W. (1982). A numerical case study of the effects of latent heating on a developing wave cyclone. *J. Atmos. Sci.*, **39**, 1555–1570.

Chang, C. B., Perkey, D. J., and Kreitzberg, C. W. (1984). Latent heat induced energy transformations during cyclogenesis. *Mon. Wea. Rev.*, **112**, 357–367.

Chang H., Kumar, A., Niyogi, D., *et al.* (2009). The role of land surface processes on the mesoscale simulation of the July 26, 2005 heavy rain event over Mumbai, India. *Glob. Planet. Change*, doi:10.1016/j.gloplacha.2008.12.005.

Changnon, S. A., Semonin, R. G., Auer, A. H., Braham, R. R., and Hales, J. M. (Eds.) (1981). *METROMEX: A Review and Summary*. Meteorological Monograph, No. 18, American Meteorological Society.

Charney, J., Quirk, W. J., Chow, S. H., and Kornfield, J. (1977). A comparative study of the effects of albedo change on drought in semi-arid regions. *J. Atmos. Sci.*, **34**(9), 1366–1401.

Charney, J. G. (1955). The use of primitive equations of motion in numerical prediction. *Tellus*, **7**, 22–26.

Chase, T. N., Pielke Sr., R. A., Kittel, T. G. F., *et al.* (2001). The relative climatic effects of landcover change and elevated carbon dioxide combined with aerosols: A comparison of model results and observations. *J. Geophys. Res.*, **106**, 31,685–31,691.

Chen, C., and Cotton, W. R. (1983). A one-dimensional simulation of the stratocumulus capped mixed layer. *Bound. Layer Meteorol.*, **25**, 298–321.

Chen, D.-X., and Coughenour, M. B. (1994). GEMTM: A general model for energy and mass transfer of land surfaces and its application at the FIFE sites. *Agr. Forest Meteorol.*, **68**, 145–171.

Chen, D.-X., and Coughenour, M. B. (2004). Photosynthesis, transpiration, and primary productivity: Scaling up from leaves to canopies and regions using process models and remotely sensed data. *Global Biogeochem. Cycles*, **18**, GB4033, doi:10.1029/2002GB001979.

Chen, D.-X., Coughenour, M. B., Knapp, A. K., and Owensby, C. E. (1994). Mathematical simulation of C4 grass photosynthesis in ambient and elevated CO_2. *Ecol. Model.*, **73**, 63–80.

Chen, D.-X., Hunt, H. W., and Morgan, J. A. (1996). Responses of a C3 and C4 perennial grass to CO_2 enrichment and climate change: Comparison between model predictions and experimental data. *Ecol. Model.*, **87**, 11–27.

Claussen, M. (2001). Biogeophysical feedbacks and the dynamics of climate. In E. D. Schulze, S. P. Harrison, M. Heimann, *et al.* (eds.), *Global Biogeochemical Cycles in the Climate System*. San Diego: Academic Press, pp. 61–71.

Claussen, M. (ed.) (2004). Does land surface matter in climate and weather? In P. Kabat, M. Claussen, P. Dirmeyer, *et al.* (eds.), *Vegetation, Water, Humans and the Climate, Part A.* Heidelberg: Springer, pp. 5–154.

Claussen, M. (2009). Late Quaternary vegetation–climate feedbacks. *Clim. Past*, **5**, 203–216.

Claussen, M., and Gayler, V. (1997). The greening of Sahara during the mid-Holocene: Results of an interactive atmosphere – biome model. *Global Ecol. Biogeog. Lett.*, **6**, 369–377.

Collatz, J., Ball, J. T., Grivet, C., and Berry, J. (1991). Physiological and environmental regulation of stomatal conductance, photosynthesis and transpiration: A model that includes a laminar boundary layer. *Agric. For. Meteorol.*, **54**, 107–136.

Collatz, J., Ribas-Carbo, M., and Berry, J. (1992). Coupled photosynthesis–stomatal conductance model for leaves of C4 plants. *Aust. J. Plant Physiol.*, **19**, 519–538.

Cotton, W. R., Weaver, J. F., and Beitler, B. A. (1995). An unusual summertime downslope wind event in Fort Collins, Colorado on 3 July 1993. *Wea. Forecasting*, **10**, 786–797.

Cotton, W. R., Pielke Sr., R. A., Walko, R. L., *et al.* (2003). RAMS 2001: current status and future directions. *Meteor. Atmos. Phys.*, **82**, 5–29.

Coulter, G. W. and Spigel, R. H. (1991). Hydrodynamics. In G. W. Coulter (ed.), *Lake Tanganyika and its Life*, New York: Oxford University Press, pp. 49–75.

Council for Agricultural Science and Technology (2004). *Climate Change and Greenhouse Gas Mitigation: Challenges and Opportunities for Agriculture.* Ames, Iowa, USA.

Crook, N. A. (1996). Sensitivity of moist convection forced by boundary layer processes to low-level thermodynamics fields. *Mon. Wea. Rev.*, **124**, 1767–1785.

Crucifix, M., and Loutre, M. F. (2002). Transient simulations over the last interglacial period (126–115 kyr BP): Feedback and forcing analysis. *Clim. Dyn.*, **19**, 417–433, doi:10.1007/s00382-002-0234-z).

Crucifix, M., Loutre, M.-F., Tulkens, P., Fichefet, T., and Berger, A. (2002). Climate evolution during the Holocene: A study with an Earth system model of intermediate complexity. *Clim. Dyn.*, **19**, 43–60

Csanady, G. T. (1967). Large-scale motion in the great lakes. *J. Geophys. Res.*, **72**(16), 4151–4162.

Curtis, P. S., and Wang, X. (1998). A meta-analysis of elevated CO_2 effects on woody plant mass, form, and physiology, *Oecologia*, **113**, 299–313.

D'Alessio, S. J. D., Abdella, K., and McFarlane, N. A. (1998). A new second-order turbulence closure scheme for modeling the oceanic mixed layer. *J. Phys. Oceanogr.*, **28**, 1624–1641.

Davis, C. A. (1992). Piecewise potential vorticity inversion. *J. Atmos. Sci.*, **49**, 1397–1411.

Davis, C. A., and Emanuel, K. A. (1991). Potential vorticity diagnostics of cyclogenesis. *Mon. Wea. Rev.*, **119**, 1929–1953.

Daley, R. (1991). *Atmospheric Data Analysis*, Cambridge University Press.

De Wekker, S. F. J., Steyn, D. G., Fast, J. D., Rotach, M. W., and Zhong, S. (2005). The performance of RAMS in representing the convective boundary layer structure in a very steep valley. *Environ. Fluid Mech.*, **5**, 35–62.

de Pury, D. G. G., and Farquhar, G. D. (1997). Simple scaling of photosynthesis from leaves to canopies without the errors of big-leaf models. *Plant Cell Environ.*, **20**, 537–557.

Dell'Osso, L. (1984). High-resolution experiments with the ECMWF-model: A case study. *Mon. Wea. Rev.*, **112**, 1853–1883.

Dell'Osso, L., and Rabinovich, D. (1984). A case study of cyclone development in the lee of the Alps on 18 March 1982. *Beitr. Phys. Atmos.*, **57**, 369–379.

DeMott., P. J., Sassen, K., Poellet, M. R., *et al.* (2003). African dust aerosols as atmospheric ice nuclei. *Geophys. Res. Lett.*, **30**, 1732, doi:10.1029/2003GL017410.

Descy, J.-P., Hardy, M.-A., Sténuite, S., *et al.* (2005). Phytoplankton pigments and community composition in Lake Tanganyika. *Freshwater Biol.*, **50**(4), 668–684.

Dettwiller, J., and Changnon, S. A. (1976). Possible urban effects on maximum daily rainfall at Paris, St. Louis and Chicago. *J. Appl. Meteor.*, **15**, 518–519.

Doswell, C. A. III, Brooks, H. E., and Maddox, R. A. (1996). Flash flood forecasting: An ingredients-based methodology. *Wea. Forecasting*, **10**, 560–581.

Doswell, C. A. III, Ramis, C., Romero, R. and Alonso, S. (1998). A diagnostic study of three heavy precipitation episodes in the western Mediterranean region. *Wea. Forecasting*, **13**, 102–124.

Douglas, E. M., Niyogi, D., Frolking, S., *et al.* (2006). Changes in moisture and energy fluxes due to agricultural land use and irrigation in the Indian Monsoon Belt. *Geophys. Res. Lett.*, **33**, L14403, doi:10.1029/2006GL026550.

Douglas, E. M., Beltrán-Przekurat, A., Niyogi, D. S., Pielke, Sr., R. A., and Vörösmarty, C. J. (2009). The impact of agricultural intensification and irrigation on land-atmosphere interactions and Indian Monsoon precipitation: A mesoscale modeling perspective. *Glob. Planet. Change*, doi:10.1016/j.gloplacha.2008.12.007.

Dudhia, J. (1989). Numerical study of convection observed during the Winter Monsoon Experiment using a mesoscale two-dimensional model. *J. Atmos. Sci.*, **46**, 3077–3107.

Eagan, R. C., Hobbs, P. V., and Radke, L. F. (1974). Particle emissions from a large Kraft paper mill and their effects on the microstructure of warm clouds. *J. Appl. Meteor.*, **13**, 535–552.

Eastman, J. L., Coughenour, M. B., and Pielke, R. A. (2001). The effects of CO_2 and landscape change using a coupled plant and meteorological model. *Glob. Change Biol.*, **7**, 797–815.

Efron, B., and Tibshirani, R. J. (1993). *An Introduction to the Bootstrap*. London: Chapman and Hall.

Egger, J. (1972). Numerical experiments on the cyclogenesis in the Gulf Genoa. *Beitr. Phys. Atmos.*, **45**, 320–346.

Egger, J. (1988) Alpine lee cyclogenesis: Verification of theories. *J. Atmos. Sci*, **45**, 2187–2203.

Ek, M. B., Mitchell, K. E., Lin, Y., *et al.* (2003). Implementation of Noah land-surface model advances in the National Centers for Environmental Prediction operational mesoscale Eta model. *J. Geophys. Res.*, 108D, 8851–8867.

Emeis, S., and Hantel, M. (1984). ALPEX diagnostics: Subsynoptic heat fluxes. *Beitr. Phys. Atmos.*, **57**, 496–511.

Ertel, H. (1942). Ein neuer hydrodynamischer wirbelsatz. *Meteor. Z.*, **59**, 271–281.

Evensen, G. (1994). Sequential data assimilation with a nonlinear quasi-geostrophic model using Monte Carlo methods to forecast error statistics. *J. Geophys. Res.*, **99**, 10,143–10,162.

Fall, S., Niyogi, D., Pielke Sr., R. A., *et al.* (2010) Impacts of land use land cover on temperature trends over the continental United States: Assessment using the North American Regional Reanalysis, *Int. J. Climatol.*, doi: 10.1002/joc.1996.

Farquhar, G., von Caemmerer, S., and Berry, J. (1980). A biochemical model of photosynthetic CO_2 assimilation in leaves of C3 species. *Planta*, **149**, 78–90.

Feingold, G., Cotton, W. R., Kreidenweis, S. M., and Davis, J. T. (1999). The impact of giant cloud condensation nuclei on drizzle formation in stratocumulus: Implications for cloud radiative properties. *J. Atmos. Sci.*, **56**, 4100–4117.

Foley, J. A., DeFries, R., Asner, G. P., *et al.* (2005). Global consequences of land use. *Science*, **309**, 570–574.

Gagin, A. (1965). Ice nuclei, their physical characteristics and possible effect on precipitation initiation. *Proc. Int. Conf. on Cloud Physics*, Tokyo and Sapporo, Japan, pp. 155–162.

Gallée, H., van Ypersele, J.-P., Fichefet, T., Tricot, C., and Berger, A. (1991). Simulation of the last glacial cycle by a coupled, sectorially averaged climate-ice sheet model. I. The climate model. *J. Geophys. Res.*, **96**, 13,139–13,161.

Gallée, H., van Ypersele, J.-P., and Fichefet, T., *et al.* (1992). Simulation of the last glacial cycle by a coupled, sectorially averaged climate-ice-sheet model. II. Response to insolation and CO_2 variation. *J. Geophys. Res.*, **97**, 15,713–15,740.

Ganopolski, A., Kubatzki, C., Claussen, M., Brovkin, V., and Petoukhov, V. (1998). The influence of vegetation-atmosphere -ocean interaction on climate during the Mid-Holocene. *Science*, **280**, 5371, 1916–1919.

Garstang, M. (1967). Sensible and latent heat exchange in low-latitude synoptic-scale systems. *Tellus*, **19**, 492–509.

Gero, A., and. Pitman, A. J. (2006). The impact of land cover change on a simulated storm event in the Sydney Basin. *J. Appl. Meteor.*, **45**, 283–300.

Ghan, S. J., Leung, L. R., and McCaa, J. (1999). A comparison of three different modeling strategies for evaluating cloud and radiation parameterizations. *Mon. Wea. Rev.*, **127**, 1967–1984.

Goswami, B. N. and Shukla, J. (1984). Quasi-periodic oscillations in a symmetric general circulation model. *J. Atmos. Sci.*, **41**(1), 20–37.

Goudriaan, J. (1977). *Crop Micrometeorology: A Simulation Study*. Wageningen: Pudoc.

Gourgue, O., Deleersnijder, E., and White, L.(2007). Toward a generic method for studying water renewal, with application to the epilimnion of Lake Tanganyika. *Estuar. Coast. Shelf Sci.*, 74(4), 628–640.

Guan, S., and Reuter, G. W. (1995). Numerical simulation of a rain shower affected by waste energy released from a cooling tower complex in a calm environment. *J. Appl. Meteor.*, **34**, 131–142.

Hacker J. P., and Snyder, C. (2005). Ensemble Kalman filter assimilation of fixed screen-height observations in a parameterized PBL. *Mon. Wea. Rev.*, **133**, 3260–3275.

Hacker, J. P., and Rostkier-Edelstein, D. (2007). PBL state estimation with surface observations, a column model and an ensemble filter. *Mon. Wea. Rev.*, **135**, 2958–2972.

Hacker, J. P., and Rostkier-Edelstein, D. (2008). PBL state estimation with surface observations, a column model, and an ensemble filter: probabilistic evaluation under various mesoscale regimes *Proceedings of the 88th Annual Meeting, AMS*, New Orleans, LA, 20–24 January 2008.

Hanert, E., Le Roux, D. Y., Legat, V., and Deleersnijder, E. (2005). An efficient Eulerian finite element method for the shallow water equations. *Ocean Model.*, **10**(1–2): 115–136.

Hansen, J., Jones, J. W., Kiker, C. F., and Hodges, A. W. (1999). El Niño-Southern Oscillation impacts on winter vegetable production in Florida. *J. Climate*, **12**, 92–102.

Hansen, J., and Nazarenko, L. (2004). Soot climate forcing via snow and ice albedos. *Proc. Natl. Acad. Sci.*, **101**, 423–428, doi:10.1073/pnas.2237157100.

Harrington, J. Y. (1997). The effects of radiative and microphysical processes on simulated warm and transition season Arctic stratus. Ph.D. Dissertation, Atmospheric Science Paper No 637, Colorado State University.

Harshvardan, and Corsetti, T. G. (1984). *Long Wave Parameterization for the UCLA/GLAS GCM*. NASA Technical Memo 86072, Goddard Space Flight Center, Greenbelt, MD.

Hatfield, J. L. (2008). Climate and agriculture: Challenges for efficient production [CD-ROM]. In *Symposium of the Climate Information for Managing Risks*, June 10–13, St. Pete Beach, FL, USA.

Haurwitz, B. (1947). Comments on the sea-breeze circulation, *J. Meteor.*, **4**, 1–8,

Held, I., and Suarez, M. (1974). Simple albedo feedback models of the icecaps. *Tellus*, **36**, 613–629.

Hindman, E. E., II, Hobbs, P. V., and Radke, L. F. (1977a). Cloud condensation nuclei from a paper mill. Part I: Measured effects on clouds. *J. Appl. Meteor.*, **16**, 745–752.

Hindman, E. E. II, Tag, P. M., Silverman, B. A., and Hobbs, P. V. (1977b). Cloud condensation nuclei from a paper mill. Part II: Calculated effects on rainfall. *J. Appl. Meteor.*, **16**, 753–755.

Hobbs, P. V., Radke, L. F., and Shumway, S. E. (1970). Cloud condensation nuclei from industrial sources and their apparent influence on precipitation in Washington State. *J. Atmos. Sci.*, **27**, 81–89.

Holtslag, A. A. M., and Boville, B. A. (1993). Local versus nonlocal boundary layer diffusion in a global climate model. *J. Climate*, **6**, 1825–1847.

Homar, V., Ramis, C., Romero, R., *et al.* (1999) A case of convection development over the western Mediterranean sea: A study through numerical simulations. *Meteor. Atmos. Phys.*, **71**, 169–188.

Hong, S.-Y., and Pan, H.-L. (1996). Nonlocal boundary layer vertical diffusion in a mediumtrange forecast model. *Mon. Wea. Rev,.***124**, 2322–2339.

Hoogenboom, G., Jones, J. W., and Boote, K. J. (1992). Modeling the growth, development and yield of grain legumes using SOYGRO, PNUTGRO and BEANGRO: A review, *Trans. ASAE*, **35**, 2043–2056.

Hoogenboom, G., Jones, J. W., Wilkens, P. W., *et al.* (1994). Crop models. In G. Y. Tsuji, G. Uehara, and S. Balas (eds.), *DSSAT3, 2–2*, University of Hawaii, Honolulu, Hawaii.

Hoogenboom, G., Jones, J. W., Wilkens, P. W., *et al.* (2004). Decision Support System for Agrotechnology Transfer Version 4.0 [CD-ROM], University of Hawaii, Honolulu, Hawaii.

Idso, K. E., and Idso, S. B. (1994). Plant responses to atmospheric CO_2 enrichment in the face of environmental constraints: A review of the past ten year's research. *Agric. Forest Meteorol.* **69**, 153–203.

Hoskins, B. J, McIntyre, M. E., and Robertson, A. W. (1985). On the use and significance of isentropic potential vorticity maps. *Quart. J. Roy. Meteor. Soc.*, **111**, 877–946.

Houghton, J., Filho, L., Callander, B., Harris, N., and Maskell, K. (1996). *Climate Change: the Science of Climate Change*. Cambridge University Press.

Huff, F. A., and Vogel, J. L. (1978). Urban, topographic and diurnal effects on rainfall in the St. Louis Region. *J. Appl. Meteor.*, **17**, 565–577.

Hung, H.-M., Malinowski, A., and Martin, S. (2003). Kinetics of heterogeneous ice nucleation on the surfaces of mineral dust cores inserted into aqueous sulfate particles. *J. Phys. Chem. A*, **107**, 1296–1306.

Huo, Z., Zhang, D. L., and Gyakum, J. R. (1999). Interaction of potential vorticity anomalies in extratropical cyclogenesis. Part I: Static piecewise inversion. *Mon. Wea. Rev.*, **127**, 2546–2561.

Iglesias, A., Erda, L., and Rosenzweig, C. (1996). Climate change in Asia: A review of the vulnerability and adaptation of crop production. *Water, Air, Soil Poll.* **92**, 13–27.

Intergovernmental Panel on Climate Change (IPCC) (2007). *Climate Change 2007: The Scientific Basis. Summary for Policy Makers*. Cambridge: Cambridge University Press.

Isono, K., Komabayasi, M., and Ono, A. (1959). The nature and origin of ice nuclei in the atmosphere. *J. Meteorol. Soc. Japan*, **37**, 211–233.

Iversen, T., and Nordeng, T. E. (1984). A hierarchy of nonlinear filtered models: Numerical solutions. *Mon. Wea. Rev.*, **112**, 2048–2059.

Jablonski, L. M., Wang, X., and Curtis, P. S. (2002). Plant reproduction under elevated CO_2 conditions: A meta-analysis of reports on 79 crop and wild species, *New Phytologist*, **156**, 9–26.

Jahn, A., Claussen, M., Ganopolski, A., and Brovkin, V. (2005). Quantifying the effect of vegetation dynamics on the climate of the Last Glacial Maximum. *Climate of the Past*, **1**, 1–7.

Jansà, A., Genovés, A., Picornell, M. A., *et al.* (2001). Western Mediterranean cyclones and heavy rain. Part 2: Statistical approach. *Meteorol. Appl.*, **8**, 43–56.

Jeffreys, H. (1922). On the dynamics of wind. *Quart. J. Roy. Meteor. Soc.*, **48**, 29–47.

Jensen, E., Starr, D., and Toon, B. (2004). Mission investigates tropical cirrus clouds. *EOS, Trans. Amer. Geophys. Union*, **84**, 45–50.

Jolliffe, I. T. and Stephenson, D. B. (Eds.) (2003). *Forecast Verification, A Practitioner's Guide in Atmopsheric Science*. New York: Wiley.

Jones, C. A., and Kiniry, J. R. (1986). *CERES-Maize: A Simulation Model of Maize Growth and Development*. College Station, Texas: Texas A&M University Press.

Jones, J. W., Hoogenboom, G., Porter, C. H., *et al.* (2003). DSSAT Cropping System Model. *Eur. J. Agronomy*, **18**, 235–265.

Kain, J. S. (2004). The Kain–Fritsch convective parameterization: An update. *J. Appl. Meteor.*, **43**, 170–181.

Kalnay, E. (2003). *Atmospheric Modeling, Data Assimilation and Predictability*, Cambridge University Press.

Kalnay, M., Kanamitsu, M., Kistler, R., *et al.* (1996). The NCEP/NCAR 40-Year reanalysis project. *Bull. Am. Meteor. Soc.*, **77**, 437–471.

Kanamitsu, M. (1975). *On Numerical Prediction Over a Global Tropical Belt*. Department of Meteorology Report 75-1, The Florida State University, Tallahassee.

Kanamitsu, M. (1989). Description of the NMC global data assimilation and forecast system. *Wea. Forecast.*, **4**, 335–342.

Kanamitsu, M., Tada, K., Kudo, K., Sata, N., and Isa, S. (1983). Description of the JMA operational model. *J. Meteor. Soc. Japan*, **61**, 812–828.

Kanari, S. (1975). The long-period internal waves in Lake Biwa. *Limnol. Oceanogr.*, **20**(4), 544–553.

Kaufman, Y. J., and Nakajima, T. (1993). Effect of Amazon smoke on cloud microphysics and albedo-analysis from satellite imagery. *J. Appl. Meteor.*, **32**, 729–744.

Khain, A. P., Rosenfeld, D., and Sednev, I. (1993). Coastal effects in the E. Mediterranean as seen from experiments using a cloud ensemble model with detailed description of warm and ice microphysical processes. *Atmos. Res.*, **30**, 295–319.

Kenney, S. E., and Smith, P. J. (1983). On the release of eddy available potential energy in an extratropical cyclone system. *Mon. Wea. Rev.*, **111**, 745–755.

Kielmann, J., and Simons, T. J. (1984). Some aspects of baroclinic circulation models. In K. Hutter (ed.), *Hydrodynamics of Lakes: CISM Lectures*, New York, Vienna: Springer Verlag, pp. 235–285.

Kitade, T. (1983). Nonlinear normal mode initialization with physics. *Mon. Wea. Rev.*, **111**, 2194–2213.

Krichak, S. O., and Alpert, P. (2002). A fractional approach to the factor separation method. *J. Atmos. Sci.*, **59**, 2243–2252.

Krichak. S., Alpert, P., and Krishnamurti, T. N. (1997). Red Sea trough/cyclone development numerical investigation. *Meteor Atmos. Phys.*, **63**, 159–170.

Krichak. S., and Alpert, P. (1998). Role of the large-scale moist dynamics in Nov 1–5 1994 hazardous Mediterranean weather. *J. Geophy. Res.*, **103**, 19,453–19,468.

Krishnamurti, T. N. (1968). A diagnostic balanced model for studies of weather systems of low and high latitudes, Rossby number less than 1. *Mon. Wea. Rev.*, **96**, 197–207.

Krishnamurti, T. N. and Bhalme, H. N. (1976). Oscillations of a Monsoon system. Part I. Observational aspects. *J. Atmos. Sci.*, **33**(10), 1937–1954.

Krishnamurti, T. N., Xue, J., Bedi, H. S., Ingles, K. and Oosterhof, D. (1991). Physical initialization for numerical weather prediction over the tropics. *Tellus*, 43AB, 53–81.

Krishnamurti, T. N., Bedi, H. S., Rohaly, G. D., and Oosterhof, D. (1996). Partitioning of the seasonal simulation of a monsoon climate. *Mon. Wea. Rev.*, **124**, 1499–1519.

Krishnamurti, T. N., Wagner, C. P., Cartwright, T. J., and Oosterhof, D. (1997). Wave trains excited by cross-equatorial passage of the monsoon annual cycle. *Mon. Wea. Rev.*, **125**, 2709–2715.

Krishnamurti, T. N., Sanjay, J., Mitra, A. K., and Vijaya Kumar, T. S. V. (2004). Determination of forecast errors arising from different components of a model physics and dynamics. *Mon. Wea. Rev.*, **132**, 2570–2594.

Krishnamurti, T. N., Basu, S., Sanjay, J., and Gnanaseelan, C. (2008). Evaluation of several different planetary boundary layers within a single model, a unified model and a multimodel superensemble. *Tellus*, **60A**, 42–61.

Kubatzki, C., Claussen, M. (1998). Simulation of the global bio-geophysical interactions during the Last Glacial Maximum. *Clim. Dyn.*, **14**, 461–471.

Kubatzki, C., Montoya, M., Rahmstorf, S., Ganopolski, A., and Claussen, M. (2000). Comparison of a coupled global model of intermediate complexity and an AOGCM for the last interglacial. *Clim. Dyn.*, **16**, 799–814.

Küchler, A. W. (2000). World map of natural vegetation. *Goode's World Atlas*, 20th edn. Rand McNally, pp. 24–25.

Kumar, N. and Russell, A. G. (1996). Comparing prognostic and diagnostic meteorological fields and their impacts on photochemical air quality modeling. *Atmos. Environ.*, **12**, 1989–2010.

Kumar, V., Bhagat, D. K., Kumar, U. R., and Ganesh, M. S. (2007). Impact of low level jet on heavy rainfall events over Mumbai, *Mausam*, **58**, 229–240.

Kuo, Y.-H., and Low-Nam, S. (1990). Prediction of nine explosive cyclones over the western Atlantic Ocean with a regional model. *Mon. Wea. Rev.*, **118**, 3–25.

Kutzbach, J., Bonan, G., Foley, J., Harrison, S. P. (1996). Vegetation and soil feedback on the response of the African monsoon to orbital forcing in the early to middle Holocene. *Nature*, **384**, 623–626.

Lacis, A. A., and Hansen, J. E. (1974). A parameterization for the absorption of solar radiation in the Earth's atmosphere. *J. Atmos. Sci.*, **31**, 118–133.

Laird, N. F., Ochs III, H. T., Rauber, R. M., and Miller, L. J. (2000). Initial precipitation formation in warm Florida cumulus. *J. Atmos. Sci.*, **57**, 3740–3751.

Lal, M., Singh, K. K., Srinivasan, G., et al. (1998). Growth and yield responses of soybean in Madhya Pradesh, India, to climate variability and change. *Agric. Forest Meteorol.*, **93**, 53–70.

Lannici, J. M., Carlson, T. N., and Warner, T. T. (1987). Sensitivity of the Great Plains severe storm environment to soil-moisture distribution. *Mon. Wea. Rev.*, **115**, 1005–1016.

Lau, K. M. (1992). East Asian summer monsoon variability and climate teleconnection. *J. Meteor. Soc. Japan*, **70**, 211–242.

Lawlor, D. W., and Mitchell, R. A. C. (1991). The effects of increasing CO_2 on crop photosynthesis and productivity: A review of field studies. *Plant, Cell Environ.*, **14**, 807–818.

Lawton, R. O., Nair, U. S., Pielke Sr., R. A., and Welch, R. M. (2001). Climatic impact of tropical lowland deforestation on nearby montane cloud forests. *Science*, **294**, 584–587.

L'Ecuyer, T. S., Berg, W., Haynes, J., Lebsock, M., and Takemura, T. (2009). Global observations of aerosol impacts on precipitation initiation in warm-topped maritime clouds. *J. Geophys. Res.*, **114**, D09211.

Lei, M., Niyogi, D., Kishtawal, C., *et al.* (2008). Effect of explicit urban land surface representation on the simulation of the 26 July 2005 heavy rain event over Mumbai, India. *Atmos. Chem. Phys. Discuss.*, **8**, 8773–8816.

Leidner, S. M., Stauffer, D. R., and Seaman, N. (2001) Improving short-term numerical weather prediction in the California coastal zone by dynamic initialization of the marine boundary layer, *Mon. Wea. Rev.*, **129**, 275–294.

Leslie, L., Holland, G. J., and Lynch, A. H. (1987). Australian east coast cyclones: Numerical modeling study. *Mon. Wea. Rev.*, **115**, 3037–3053.

Levi, Y., and Rosenfeld, D. (1996). Ice nuclei, rainwater chemical composition, and static cloud seeding effects in Israel. *J. Appl. Meteor.*, **35**, 1494–1501.

Levy, G., and Ek, M. B. (2001). The simulated response of the marine atmospheric boundary layer in the western pacific warm pool region to surface forcing. *J. Geophys. Res.*, **106**, 7229–7241.

Levin, Z., Ganor, E., and Gladstein, V. (1996). The effects of desert particles coated with sulfate on rain formation in the Eastern Mediterranean. *J. Appl. Meteor.*, **35**, 1511–1523.

Lin, J. W.-B., Neelin, J. D., and Zeng, N. (2000). Maintenance of tropical intraseasonal variability: Impact of evaporation-wind feedback and midlatitude storms. *J. Atmos. Sci.*, **57**(17), 2793–2823.

Lindzen, R. S., and Farrell, B. (1977). Some realistic modifications of simple climate models. *J. Atmos. Sci.*, **34**, 1487–1501.

Lindzen, R. S. (1990). *Dynamics in Atmospheric Physics*. Cambridge University Press.

Lizaso, J. I., Batchelor, W. D., Boote, K. J., and Westgate, M. E. (2005). Development of a leaf-level canopy assimilation model for CERES-Maize. *Agron. J.*, **97**, 722–733.

Lorenz, E. N. (1963). Deterministic nonperiodic flow, *J. Atmos. Sci.*, **20**, 130–141.

Louis, J. F. (1979). A parametric model of vertical eddy fluxes in the atmosphere. *Bound.-Layer Meteor.*, **17**, 187–202.

Lu, L., and Shuttleworth, W.J. (2002). Incorporating NDVI-derived LAI into the climate version of RAMS and its impact on regional climate. *J. Hydrometeor.*, **3**, 347–362.

Lynn, B. H., Stauffer, D. R., Wetzel, P. J. *et al.* (2001). Improved simulation of Florida summertime convection using the PLACE land-surface model and a 1.5-order turbulence parameterization coupled to the Penn State/NCAR mesoscale model. *Mon. Wea. Rev.*, **129**, 1441–1461.

Madden, R. A., and Julian, P. R. (1971). Detection of 40–50 day oscillation in the zonal wind in the Tropical Pacific. *J. Atmos. Sci.*, **28**(5), 702–708.

Madden, R. A., and Julian, P. R.(1994). Observations of the 40–50-day tropical oscillation: A review. *Mon. Wea. Rev.*, **122**(5), 814–837.

Maddox, R. A., Perkey, D. J., and Fritsch, J. M. (1981). Evolution of upper troppospheric features during the development of a mesoscale convective complex. *J. Atmos. Sci.*, **38**, 1664–1674.

Mahrer, Y., and Pielke, R. A. (1977). A numerical study of the airflow over irregular terrain. *Beitr. Phys. Atmos.*, **50**, 98–113.

Magrín, G. O., Travasso, M. I., and Rodríguez, G. R. (2002). *Changes in Climate and Maize Production During the 20th Century in Argentina*. INTA, Instituto de Clima y Agua, 1712 Castelar, Argentina.

Mailhot, J., and Chouinard, C. (1989). Numerical forecasts of explosive winter storms: Sensitivity experiments with a meso-α scale model. *Mon. Wea. Rev.*, **117**, 1311–1343.

Mall, R. K., Lal, M., Bhatia, V.S., Rathore, L.S., and Singh, R. (2004). Mitigating climate change impact on soybean productivity in India: A simulation study. *Agric. Forest Meteorol.* **121**, 113–125.

Manobianco, J. (1988). On the observational and numerical aspects of explosive East Coast cyclogenesis. Ph.D. Thesis, The Florida State University, USA.

Marchal, E. (2005). Simulation par lments finis des oscillations induites par le vent de la thermocline dans le Lac Tanganyika. Master's thesis, Université catholique de Louvain.

Martin, W. J., and Xue, M. (2006). Sensitivity analysis of convection of the 24 May 2002 IHOP case using very large ensembles. *Mon. Wea. Rev.*, **134**, 192–207.

Mason, I. (1982). A model for assessment of weather forecasts. *Aust. Meteorol. Mag.*, **30**, 291–303.

Masson, V. (2000). A physically-based scheme for the urban energy budget in atmospheric models. *Bound. Layer Meteor.*, **94**, 357–397.

Mason, S. J., and Graham, N. E. (1999). Conditional probabilities, relative operating characteristics, and relative operative levels. *Wea. Forecasting*, **14**, 713–725.

Mather, G. K. (1991). Coalescence enhancement in large multicell storms caused by the emissions from a Kraft paper mill. *J. Appl. Meteor.*, **30**, 1134–1146.

Matthews, E. (1983). Global vegetation and land use: New high-resolution data bases for climate studies. *J. Clim. Appl. Meteor.*, **22,** 474–487.

Mauritsen T., Svensson, G., Zilitinkevich, S. S., *et al.* (2007). A total turbulent energy closure model for neutrally and stably stratified atmospheric boundary layers. *J. Atmos. Sci.*, **64**, 4113–4126.

McCaul, E. W., and Cohen, C. (2002). The impact on simulated storm structure and intensity of variations in the mixed layer and moist layer depths. *Mon. Wea. Rev.*, **130**, 1722–1748.

McGinley, J. A., and Goerss, J. S. (1986). Effects of terrain height and blocking initialization on numerical simulation of Alpine lee cyclogenesis. *Mon. Wea. Rev.*, **117**, 1311–1343.

Mellor, G.L., and Yamada, T. (1974). A hierarchy of turbulence closure models for planetary boundary layers. *J. Atmos. Sci.*, **31**, 1791–1806.

Mellor, G. L., and Yamada, T. (1982). Development of a turbulence closure model for geophysical fluid problems. *Rev. Geophys. Space Phys.*, **20**, 851–875.

Mera, R. J., Niyogi, D., Buol, G. S., Wilkerson, G. G., and Semazzi, F. H. M. (2006). Potential individual versus simultaneous climate change effects on soybean (C3) and maize (C4) crops: An agrotechnology model based study. *Glob. Planet. Change*, **54**, 163–182.

Mera, R. J., Niyogi, D. S., Booker, F. L., *et al.* (2010). Soybean sensitivity to simulated changes in rainfall, temperature, radiation and irrigation at ambient and elevated carbon dioxide concentrations. *J. Appl. Meteor. Climatol.*, submitted.

Mesinger, F., and Strickler, F. (1982). Effects of mountains on Genoa cyclogenesis. *J. Meteor. Soc. Japan*, **60**, 326–338.

Meteorological Office (1962). Weather in the Mediterranean. Air Ministry, Vol. 1.

Meyers, M. P., DeMott, P. J., and Cotton, W. R. (1992). New primary ice nucleation parameterizations in an explicit cloud model. *J. Appl. Meteor.*, **31**, 708–721.

Meyers, M. P., Walko, R. L., Harrington, J. Y., and Cotton, W. R. (1997). New RAMS cloud microphysics parameterization. Part II: The two-moment scheme. *Atmos. Res.*, **45**, 3–39.

Mlawer, E. J., Toubman, S. J., Brown, P. D., Iacono, M. J., and Clough, S. A. (1997). Radiative transfer for inhomogeneous atmosphere: RRTM, a validated correlated-k model for the long-wave. *J. Geophy., Res.*, **102D**, 16663–16682.

Molinari, J. (1985). A general form of Kuo's cumulus parameterisation. *Mon. Wea. Rev.*, **113**, 1411–1416.

Montgomery D. C. (2001). *Design and Analysis of Experiments*, 5th edn. New York: J. Wiley & Sons.

Mortimer, C. H. (1974). Lake hydrodynamics. *Mitteilungen Internationale Vereinigung für Theoretische und Angewandte Limnologie*, **20**, 124–197.

Mullen, S. L., and Baumhafner, D. P. (1988). Sensitivity of numerical simulations of explosive oceanic cyclogenesis to changes in physical parametrizations. *Mon. Wea. Rev.*, **116**, 2289–2329.

Müller, M. D., Schmutz, C., and Parlow, E. (2007). A one-dimensional ensemble forecast and assimilation system for fog prediction. *Pure Appl. Geophys.*, **164**, 1241–1264, doi:10.1007/s00024–007–0217–4.

Murphy, A. H. (1973). A new vector partition of the probability score. *J. Appl. Meteor.*, **12**, 595–600.

Murray, F. W., Koenig, L. R., and P. M. Tag, P. M. (1978). Numerical simulation of an industrial cumulus and comparison with observations. *J. Appl. Meteor.*, **17**, 655–668.

Naithani, J. and Deleersnijder, E. (2004). Are there internal Kelvin waves in Lake Tanganyika? *Geophys. Res. Lett.*, **31**(6), L06303, doi:10.1029/2003GL019156.

Naithani, J., Deleersnijder, E., and Plisnier, P.-D. (2002). Origin of intraseasonal variability in Lake Tanganyika. *Geophys. Res. Lett.*, **29**(23), 2093, doi:10.1029/2002GL015843.

Naithani, J., Deleersnijder, E., and Plisnier, P.-D.(2003). Analysis of wind-induced thermocline oscillations of Lake Tanganyika. *Environ. Fluid Mech.*, **3**(1), 23–39.

Narisma, G. T., and Pitman, A.J.(2004). The effect of including biospheric responses to CO_2 on the impact of land-cover change over Australia. *Earth Interactions*, **8**(5), 1–28.

Narisma G. T. and Pitman, A. J. (2006). Exploring the sensitivity of the Australian climate to regional land-cover change scenarios under increasing CO_2 concentrations and warmer temperatures. *Earth Interactions*, **10**, 1–27, doi:10.1175/EI154.1

Neelin, J. D., Held, I. M., and Cook, K. H. (1987). Evaporation-wind feedback and low-frequency variability in the tropical atmosphere. *J. Atmos. Sci.*, **44**(16), 2341–2348.

Nitta, T. (1986). Long term variations of cloud amount in the western Pacific region. *J. Meteor. Soc. Japan*, **64**, 373–390.

Nitta, T. (1987). Convective activities in the tropical western Pacific and their impact on the Northern Hemisphere circulation. *J. Meteor. Soc. Japan*, **65**, 373–390.

Niyogi D., and Raman S. (1997). Comparison of four different stomatal resistance schemes using FIFE observations. *J. Appl. Meteor.*, **36**, 903–917.

Niyogi, D., andXue, Y., (2006). Soil moisture regulates the biological response of elevated atmospheric CO_2 concentrations in a coupled atmosphere biosphere model, *Glob. Planet. Change*, **54**, Special Issue: Land-use/land-cover change and its impact on climate, 94–108.

Niyogi, D. S., Alapaty, K., and Raman, S. (1999). Uncertainty in specification of surface characteristics, Part 2: Hierarchy of interaction explicit statistical analysis. *Bound. Layer Meteor.*, **91**, 341–366.

Niyogi D., Chang, H., Saxena, V. K., *et al.* (2004). Direct observations of the effects of aerosol loading on net ecosystem CO_2 exchanges over different landscapes. *Geophys. Res. Lett.*, **31**, L20506, doi:10.1029/2004GL020915.

Niyogi, D., Alapaty, K., Raman, S, and Chen, F. (2009). Development and evaluation of a coupled photosynthesis-based gas exchange evapotranspiration model (GEM) for mesoscale weather forecasting applications. *J. Appl. Meteor. Climatol.*, **48**, 349–368.

Noilhan, J., and Planton, S. (1989). A simple parameterization of land surface processes for meteorological models. *Mon. Wea. Rev.*, **117**, 536–549.

O'Reilly, C. M., Alin, S. R., Plisnier, P.-D., Cohen, A. S., and McKee, B. A. (2003). Climate change decreases aquatic ecosytem productivity of Lake Tanganyika, Africa. *Nature*, **424**, 766–768.

Orlansky, I., and Katzfey, J. J. (1987). Sensitivity of model simulations for a coastal cyclone. *Mon. Wea. Rev.*, **115**, 2792–2821.

Orville, H. D., Eckhoff, P. A., Peak, J. E.,. Hirsch, J. H., and Kopp, F. J. (1981). Numerical simulation of the effects of cooling tower complexes on clouds and severe storms. *Atmos. Environ.*, **15**, 823–836.

O'Shay, A. J., and Krishnamurti, T. N. (2004). An examination of a model's components during tropical cyclone recurvature. *Mon. Wea. Rev.*, **132**, 1143–1166.

Otterman, J., Chou, M. D., and Arking, A. (1984). Effects of nontropical forest cover on climate. *J. Appl. Meteor.*, **23**, 762–767.

Otto, J., Raddatz, T., Claussen, M., Brovkin, V., and Gayler, V. (2009a). Separation of atmosphere-ocean-vegetation feedbacks and synergies for mid-Holocene climate. *Geophys. Res. Lett.*, **36**, L09701, doi:10.1029/2009GL037482.

Otto, J., Raddatz, T., and Claussen, M. (2009b). Climate variability-induced uncertainty in mid-Holocene atmosphere-ocean-vegetation feedbacks. *Geophys. Res. Lett.*, **36**, L23710, doi: 10.1029/2009GL041457.

Owensby, C. E, Ham, J. M, Knapp, A. K, Bremer, D, and Auen, L. M. (1997). Water vapour fluxes and their impact under elevated CO_2 in a C4-tallgrass prairie. *Glob. Change Biol.* **3**, 189–195, doi:10.1046/j.1365–2486.1997.00084.x.

Pan, H.-L., and Wu, W.-S. (1995). *Implementing a Mass Flux Convective Parameterization Package for the NMC Medium-Range Forecast Model*. NMC Office Note 409.

Peixoto, J. P., and Oort, A. H. (1992). *Physics of Climate*. New York: American Institute of Physics.

Pettersen, S. (1956). *Weather Analysis and Forecasting*. New York: MacGraw Hill.

Pielke, R. A. Sr. (1984). *Mesoscale Meteorological Modeling*. New York: Academic Press.

Pielke, R. A. Sr. (2001). Influence of the spatial distribution of vegetation and soils on the prediction of cumulus convective rainfall. *Rev. Geophys.*, **39**, 151–177.

Pielke, R. A. Sr., and Niyogi, D. (2008). The role of landscape processes within the climate system. In J. C. Otto and R. Dikaum (eds.), *Landform: Structure, Evolution, Process Control. Proceedings of the International Symposium on Landforms*. Research Training Group 437. Lecture Notes in Earth Sciences, Springer, Vol. 115.

Pielke, R. A. Sr., and Pearce, R. P. (eds.) (1994). *Mesoscale Modeling of the Atmosphere*. American Meteorological Society Monograph, Volume 25.

Pielke, R. A. Sr., Cotton, W. R., Walko, R. L., *et al.* (1992). A comprehensive meteorological modeling system: RAMS. *Meteor. Atmos. Phys.*, **49**, 69–91.

Pielke, R. A. Sr., Marland, G., Betts, R. A., *et al.* (2002). The influence of land-use change and landscape dynamics on the climate system: Relevance to climate change policy beyond the radiative effect of greenhouse gases. *Phil. Trans. A. Special Theme Issue*, **360**, 1705–1719.

Pielke, R. A. Sr., Adegoke, J., Beltrán-Przekurat, A., *et al.* (2007a). An overview of regional land use and land cover impacts on rainfall. *Tellus B*, **59**, 587–601.

Pielke, R. A. Sr., Adegoke, J. O., Chase, T. N., *et al.* (2007b). A new paradigm for assessing the role of agriculture in the climate system and in climate change. *Agric. For. Meteor., Special Issue*, **132**, 234–254.

Pierrehumbert, R. T. (1985). A theoretical model of orographically modified cyclogenesis. *J. Atmos. Sci*, **42**, 1244–1258.

Pitman, A. J. (2003). The evolution of, and revolution in, land surface schemes designed for climate models. *Int. J. Climatol.*, **23**, 479–510.

Pitman, A. J., and Narisma, G. T. (2005). The role of land surface processes in regional climate change: A case study of future land cover change over South Western Australia. *Meteor. Atmos. Phys.*, **89**, 235–249.

Plisnier, P.-D., Chitamwebwa, D., Mwape, L., *et al.* (1999). Limnological annual cycle inferred from physical-chemical fluctuations at three stations of Lake Tanganyika. *Hydrobiologia*, **407**, 45–58.

Pritchard, S. G. (2005). In S. G. Pritchard and S. Amthor (eds.), Crops and Environmental Change: An Introduction to Effects of Global Warming, Increasing Atmospheric CO_2 and O_3 Concentrations, and Soil Salinization on Crop Physiology and Yield. New York: Food Products Press.

Prospero, J. M. (1996). Saharan dust transport over the north Atlantic Ocean and Mediterranean: An overview. In S. Guerzoni and R. Chester (eds.), *The Impact of Desert Dust Across the Mediterranean*, Norwell, MA: Kluwer Academic, pp. 133–151.

Prospero, J. M. (1999). Long-term measurements of the transport of African mineral dust to the southeastern United States: Implications for regional air quality. *J. Geophys. Res.*, **104**, 15,917–15,927.

Radinovic, D. (1987). Mediterranean Cyclones and Their Influence on the Weather and Climate. WMO, PSMP Rep. Ser. 24.

Ramis, C. (1995). Las observaciones de la atmósfera libre en Mallorca: una breve historia y algunos resultados. *Revista de Ciència*, **17**, 41–58.

Ramis. C., and Romero, R. (1995). A first numerical simulation of the development and structure of the sea breeze in the Island of Mallorca. *Ann. Geophysicae*, **13**, 981–994.

Ramis, C., Romero, R., Homar, V., Alonso, S., and Alarcón, M. (1998). Diagnosis and numerical simulation of a torrential precipitation event in Catalonia (Spain). *Meteor. Atmos. Phys.*, **69**, 1–21.

Reiter, E. (1975). Handbook for Forecasters in the Mediterranean. Part 1: General Description of the Meteorological Processes. Monterey, CA: Naval Environmental Research Facility.

Rémy, S. and Bergot, T. (2010). Ensemble Kalman filter data assimilation in a 1D numerical model used for fog forecasting. *Mon. Wea. Rev.*, **138**, 1792–1810.

Reuter, G. W., and Guan, S. (1995). Effects of industrial pollution on cumulus convection and rain showers: A numerical study. *Atmos. Environ.*, **29**, 2467–2474.

Reuter, G. W., and Yau, M. K. (1987). Mixing mechanisms in cumulus congestus clouds. Part II: Numerical simulations. *J. Atmos. Sci.*, **44**, 798–827.

Rhea, J. O. (1978). Orographic precipitation model for hydrometeorological use. Ph.D. Dissertation, Colorado State University.

Roberts, P., and Hallett, J. (1968). A laboratory study of the ice nucleating properties of some mineral particulates. *Quart. J. Roy. Meteor. Soc.*, **94**, 25–34.

Romero, R. (2001). Sensitivity of a heavy rain producing Western Mediterranean cyclone to embedded potential vorticity anomalies. *Quart. J. Roy. Meteor. Soc.*, **127**, 2559–2597.

Romero, R. (2008). A method for quantifying the impacts and interactions of potential-vorticity anomalies in extratropical cyclones. *Quart. J. Roy. Meteor. Soc.*, **134**, 385–402.

Romero, R., Ramis, C., and Alonso, S. (1997). Numerical simulation of an extreme rainfall event in Catalonia: Role of orography and evaporation from the sea. *Quart. J. Roy. Meteor. Soc.*, **123**, 537–559.

Romero, R., Ramis, C., Alonso, S., Doswell III, C. A., and Stensrud, D. J. (1998). Mesoscale model simulations of three heavy precipitation events in the western Mediterranean region. *Mon. Wea. Rev.*, **126**, 1859–1881.

Romero, R., Doswell, C. A. III and Ramis, C. (2000). Mesoscale numerical study of two cases of long-lived quasistationary convective systems over eastern Spain. *Mon. Wea. Rev.*, **128**, 3731–3751.

Rosenfeld, D. (1999). TRMM observed first direct evidence of smoke from forest fires inhibiting rainfall. *Geophys. Res. Lett.*, **26**, 3105–3108.

Rosenfeld, D. (2000). Suppression of rain and snow by urban and industrial air pollution. *Science*, **287**, 1793–1796.

Rosenfeld, D., and Lensky, I. M. (1998). Satellite-based insights into precipitation formation processes in continental and maritime convective clouds. *Bull. Am. Meteor. Soc.*, **79**, 2457–2476.

Rosenfeld, D., and Woodley, W. L. (1989). Effects of cloud seeding in west Texas. *J. Appl Meteor.*, **28**, 1050–1080.

Rosenfeld, D., and W. L. Woodley, W. L. (1993). Effects of cloud seeding in west Texas: Additional results and new insights. *J. Appl. Meteor.*, **32**, 1848–1866.

Rosenfeld, D., Lahav, R., Khain, A., and Pinsky, M. (2002). The role of sea spray in cleansing air pollution over ocean via cloud processes. *Science*, **297**, 1667–1670.

Rossby, C. G. (1940). Planetary flow patterns in the atmosphere. *Quart. J. Roy. Meteor. Soc.*, **66** (Suppl.), 68–87.

Rostkier-Edelstein, D., and Hacker, J. P. (2009). Probabilistic nowcasting of PBL profiles with surface observations and an ensemble filter. *23rd Conference on Weather Analysis and Forecasting/19th Conference on Numerical Weather Prediction*, AMS, Omaha, NE, 1–5 June 2009.

Rostkier-Edelstein, D., and Hacker, J. P. (2010). The roles of surface-observation ensemble assimilation and model complexity for nowcasting PBL profiles: A factor separation analysis. *Wea. Forecast.* doi: 10.1175/2010WAF2222435.1.

Roy, S. S., Mahmood, R., Niyogi, D., *et al.* (2007). Impacts of the agricultural Green Revolution-induced land use changes on air temperatures in India. *J. Geophys. Res.*, **112**, D21108, doi:10.1029/2007JD008834.

Rozoff, C. M., Cotton, W. R., and Adegoke, J. O. (2003). Simulation of St. Louis, Missouri, land use impacts on thunderstorms. *J. Appl. Meteor.*, **42**, 716–738.

Sassen, K., DeMott, P. J., Prospero, J. M., and Poellet, M. R. (2003). Saharan dust storms and indirect aerosol effects on clouds: CRYSTAL-FACE results. *Geophys. Res. Lett.*, **30**, 1633, doi:10.1029/2003GL017371.

Schaefer, V. J. (1949). The formation of ice crystals in the laboratory and the atmosphere. *Chem. Rev.*, **44**, 291.

Schaefer, V. J. (1954). The concentrations of ice nuclei in air passing the summit of Mt. Washington. *Bull. Amer. Meteor. Soc.*, **35**, 310–314.

Schlesinger, M. E. (1988). Quantitative analysis of feedbacks in climate model simulations of CO_2-induced warming. In M. E. Schlesinger (ed.), *Physically-based modeling and simulation of climate and climate change, Part 2*. NATO ASI Series C, Vol. 243, pp. 653–735.

Schwab, D. J. (1977). Internal free oscillations in Lake Ontario. *Limnol. Oceanog.*, **22**(4), 700–708.

Sellers, P., Berry, J., Collatz, J., Field, C., and Hall, F. (1996). Canopy reflectance, photosynthesis, and transpiration. III. A reanalysis using improved leaf models and a new canopy integration scheme. *Remote Sens. Environ.*, 187–216.

Shafran, P. C., Seaman, N. L., and Gayno, G. A. (2000). Evaluation of numerical predictions of boundary layer structure during the Lake Michigan Ozone Study. *J. Appl. Meteor.*, **39**, 412–426.

Sholokhman, T., and Alpert, P. (2007). Factor separation in atmospheric modeling: A review. In A. Ebel and T. L. Davitashvili (eds.), *Air, Water and Soil Quality Modeling for Risk and Impact Assessment*. Dordrecht: Springer, pp. 165–180.

Simpson, J., Brier, G. W., and Simpson, R. H. (1967). STORMFURY cumulus seeding experiment 1965: Statistical analysis and main results. *J. Atmos. Sci.*, **24**, 508–521.

Skamarock, W. C., Klemp, J. B., Dudhia, J. *et al.* (2005). *A description of the Advanced Research WRF Version 2*. NCAR Tech Notes 468+STR.

Smagorinsky, J. (1963). General circulation experiments with the primitive equations. Part I: The basic experiment. *Mon. Wea. Rev.*, **91**, 99–164.

Smith, R. B. (1984). A theory of lee cyclogenesis. *J. Atmos. Sci*, **41**, 1159–1168.

Solomon, S., et al. (eds.) (2007). Climate Change 2007: The Physical Science Basis. Contribution of Working Group I to the Fourth Assessment Report of the IPCC. Cambridge: Cambridge University Press.

Southworth, J., Randolph, J. C., Habeck, M., *et al.* (2000). Consequences of future climate change and changing climate variability on maize yields in the midwestern United States. *Agric. Ecosystems Environ.*, **82**, 139–158.

Stein, U., and Alpert, P. (1993). Factor separation in numerical simulations. *J. Atmos. Sci.*, **50**, 2107–2115.

Stocker, T. F. (1998). The seesaw effect. *Science*, **282**(5386), 61–62.

Strack, J. E., Pielke Sr., R. A., Steyaert, L. T., and Knox, R. G. (2008). Sensitivity of June near-surface temperatures and precipitation in the eastern United States to historical land cover changes since European settlement. *Water Resour. Res.*, **44**, W11401, doi:10.1029/2007WR00654.

Stull, R. B. (1988). *An Introduction to Boundary Layer Meteorology*. Dordrecht: Kluwer Academic Publishers.

Tafferner, A., and Egger, J. (1990). Test of theories of lee cyclogenesis. *J. Atmos. Sci.*, **47**, 2417–2428.

Talagrand, O. (1997). Assimilation of observations, an introduction. *J. Meteor. Soc. Japan, Special Issue*, **75**(1B), 191–209.

Terradellas, E., and Cano, D. (2007). Implementation of a single column model for fog and low cloud forecasting at Central-Spanish airports. *Pure Appl. Geophys.*, **164**, 1327–1345, doi:10.1007/s00024–007–0221–8.

Tibaldi, S., and Buzzi, A. (1983). Effects of orography on Mediterranean lee cyclogenesis and its relationships to European blocking. *Tellus*, **35A**, 269–286.

Tibaldi, S., Buzzi, A., and Malguzzi, P. (1980). Orographically induced cyclogenesis: Analysis of numerical experiments. *Mon. Wea. Rev.*, **108**, 1302–1314.

Tibaldi, S., Buzzi, A., and Speranza, A. (1990). Orographic cyclogenesis. In C. W. Newton and E. O. Holopainen (eds.), *Extratropical Cyclones: The Eric Palmen Memorial Volume*. American Meteorological Society, pp. 107–127.

Tiedke, M. (1984). The sensitivity of the time-mean large-scale flow to cumulus convection in the ECMWF model. *Proc. Workshop on Convection in Large-Scale Numerical Models*, Reading, Unitd Kingdom, ECMWF, pp. 297–316.

Tosi, E., Fantini, M., and Trevisan, A. (1983). Numerical experiments on orographic cyclogenesis: Relationship between the development of the lee cyclone and basic flow characteristics. *Mon. Wea. Rev.*, **111**, 799–814.

Tremback, C. J. (1990). Numerical simulation of a mesoscale convective complex: model development and numerical results. Ph.D. dissertation, Colorado State University.

Tsidulko, M., and Alpert, P. (2001) Synergism of upper-level potential vorticity and mountains in Genoa lee cyclogenesis: A numerical study. *Meteor. Atmos. Phys.*, **78**(3/4), 261–285.

Twohy, C. H., Kreidenweis, S. M., Eidhammer, T., *et al.* (2009). Saharan dust particles nucleate droplets in eastern Atlantic clouds. *Geophys. Res. Lett.*, **36**, L01807, doi:10.1029/2008GL035846.

Twomey, S. (1974). Pollution and the planetary albedo. *Atmos. Environ.*, **8**, 1251–1256.

Uccellini, L. W., Petersen, R. A., Brill, K. F., Kocin, P. J., and Tuccillo, J. J. (1987). Synergistic interactions between an upper level jet streak and diabatic processes that influence the development of low-level jet and secondary coastal cyclone. *Mon. Wea. Rev.*, **115**, 2227–2261.

Uppala, S. M., Kållberg, P. W., Simmons, A. J., *et al.* (2005). The ERA-40 re-analysis. *Quart. J. Roy. Meteor. Soc.*, **131**, 2961–3012.

van den Heever, S. C., and Cotton, W. R. (2007). Urban aerosol impacts on downwind convective storms. *J. Appl. Meteor. Climatol.*, **46**, 828–850.

van den Heever, S. C., Carrio, G. G., Cotton, W. R., DeMott, P. J,. and Prenni, A. J. (2006). Impacts of nucleating aerosol on Florida storms. Part I: Mesoscale simulations. *J. Atmos. Sci.*, **63**, 1752–1775.

Wagner, R., Antoniou, I., Pedersen, S. M., Courtney M. S., and Jørgensen, H. E. (2008). The influence of the wind speed profile on wind turbine performance measurements. *Wind Energy*, doi:10.1002/we.297.

Walko, R. L., Cotton, W. R., Meyers, M .P., and Harrington, J. Y. (1995). New RAMS cloud microphysics parameterization. Part I: The single-moment scheme. *Atmos. Res.*, **38**, 29–62.

Walko, R. L., Band, L. E., J., Baron, J., *et al.* (2000). Coupled atmosphere-biophysics-hydrology models for environmental modeling. *J. Appl. Meteor.*, **39**, 931–944.

Wallace, J. M., and Gutzler, D. S. (1981). Teleconnections in the geopotential height field during the Northern Hemisphere winter. *Mon. Wea. Rev.*, **109**, 784–812.

Wallace, J. M., Tibaldi, S., and Simmons, A. J. (1983). Reduction of systematic forecast errors in the ECMWF model through the introduction of envelope orography. *Quart. J. Roy. Meteor. Soc.*, **109**, 683–718.

Wang, F., Fraisse, C., Kitchen, N. R., and Sudduth, K. A. (2001). *Site-specific Evaluation of the CROPGRO-SOYBEAN Model on Missouri Claypan Soil.* University of Missouri, TEKTRAN, USDA-ARS.

Watterson, I., and Dix, M. (2003). Simulated changes due to global warming in daily precipitation means and extremes and their interpretation using the gamma distribution. *J. Geophys. Res.*, **108**, 4397, doi:10.1029/2002JD002928.

Wilks, D. S. (1995). *Statistical Methods in the Atmospheric Sciences: An Introduction.* Academic Press: San Diego.

Wolf, J., and van Diepen, C. A. (1995). Effects of climate change on grain maize yield potential in the European Community. *Clim. Change*, **29**, 299–331.

Zack, J. W., and Kaplan, M. L. (1987). Numerical simulations of the subsynoptic features associated with the AVE-SESAME 1 case. Part 1: The preconvective environment. *Mon. Wea. Rev.*, **115**, 2367–2394.

Zuberi, B., Bertram, A., Cassa, C., Molina, L., and Molina, M. (2002). Heterogeneous nucleation of ice in $(NH_4)2SO_4$-H_2O particles with mineral dust immersions. *Geophys. Res. Lett.*, **29**, 1504, doi:10.1029/2001GL014289.

Index